Contents

Preface

There is a world of things happening around you, most of which become knowable to you thanks to one or more of your five senses. Sensory perception happens so quickly, and so often, that it's easy to overlook how impressive a system you actually are.

Take a moment and think how many sensory events happened to you from the time you woke up to the time you began reading this book. It's likely that you can't even list all the sensory occurrences. Not only do you constantly sense the environment, but your senses also work together to compile a picture of the universe. For example, events such as people passing by, warm sun shining on your face, or observing that a cool breeze in the morning is getting warmer in the afternoon are all fine examples of your senses at work and your mind processing sensations. But how can a robot or gadget have similar input? You probably already know what makes this possible (you did buy a book on the topic): sensors.

Adding sensors to a circuit expands its capabilities just as your own senses expand your awareness and inform you about the world. Sensors provide an input for information about an environment and work much like your own senses. But sensation isn't the only issue with sensors. A component doesn't necessarily have the ability to draw conclusions when a particular event occurs. Say, for instance, that it is –5 degrees outside and you want to go for a walk; what should you wear? You know, of course, that a coat and winter clothing are in order, but a temperature sensor does not know this. It can certainly provide you with a temperature reading, but it does not make judgments or inferences about what you should wear—at least not at the component level. For sensors to matter in the same way that your own sensations *and* your reflection on these sensations matter, a level of data processing needs to occur on the sensor data. Ultimately, sensors are components that you wire so that, either through hardware or software, their data is processed —and that's what this book is about: how to wire sensors and process their data.

In the first part of this book, you'll learn how to wire up sensors to other components. The level of data processing isn't too robust at the component level, and the focus is really on just getting a sensor safely wired and teaching some of the basics. The second part of the book deals with how to process sensor data. You will learn how to easily and quickly write programs with Arduino to process sensor data, as well as how to wire and program a Raspberry Pi to support analog sensors.

In this book, you'll gain hands-on experience with some of the most useful and instructive sensors available. Among the sensors and applications in this book, you'll learn how to detect and respond to:

- Clicks and rotation with a potentiometer
- Distance with ultrasound
- Proximity with infrared sensors
- Light and dark with a photoresistor
- Temperature with a thermometer
- Relative humidity with a capacitive relative humidity sensor

What Sensors to Buy?

This book covers a number of specific sensors and components (a few are illustrated in Figure P-1). To make sourcing parts easier, we've included Appendix D, which lists a complete bill of materials for all the projects in this book.

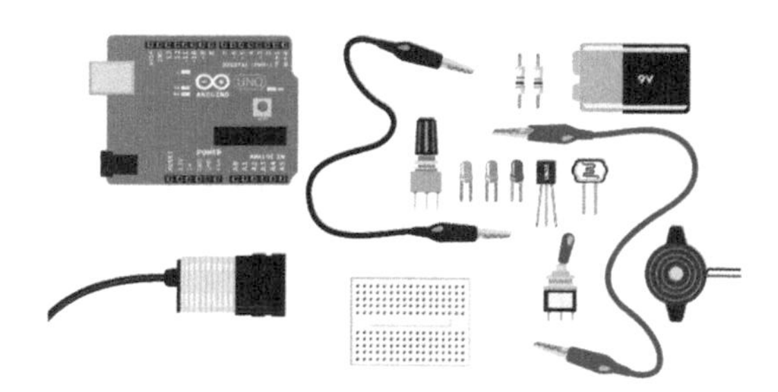

Figure P-1. *Arduino, sensors, and components*

Some well-known sellers of Arduino boards and related parts include Maker Shed (*http://makershed.com*), SparkFun Electronics (*http://www.sparkfun.com*), Parallax (*http://www.parallax.com*), and Adafruit (*http://www.adafruit.com*). All four of these shops should stock most of the individual sensors used in this book, and all sell original, high-quality parts. Start with these shops.

Global electronics distributors, like Element14 (*http://www.element14.com*) and RS Components (*http://www.rs-components.com*), are great places to order parts from, too. However, their product lists can be daunting for beginners to navigate. These global suppliers stock parts that differ from each other only in the pin format or voltage tolerances, which can be quite exhaustive. The parts sold from both of these global suppliers are high quality parts and very well documented.

Some online shops are very cheap, but these places typically do not sell official Arduino boards—they will say the product is "compatible." The sensors they sell may differ slightly in their pin configuration or even general appearance. At the time of writing, DealExtreme (*http://dx.com*) is one of the most popular shops of this sort. Even though they are based in Hong Kong and Shenzhen and offer free worldwide shipping, the quality of their parts varies a lot and delivery time can be slow. AliExpress (*http://www.aliexpress.com*) is another popular Asian shop.

If you're ordering from abroad, research your local laws regarding custom fees. In some countries, small orders may be exempt from customs and taxes.

Conventions Used in This Book

The following typographical conventions are used in this book:

Italic
: Indicates new terms, URLs, email addresses, filenames, and file extensions.

`Constant width`
: Used for program listings, as well as within paragraphs to refer to program elements such as variable or function names, databases, data types, environment variables, statements, and keywords.

`Constant width bold`
: Shows commands or other text that should be typed literally by the user.

`Constant width italic`
: Shows text that should be replaced with user-supplied values or by values determined by context.

This element signifies a tip, suggestion, or general note.

This element indicates a warning or caution.

Using Code Examples

You can download all the source code for this book from *http://getstarted.botbook.com*.

You can extract the zip package by double-clicking it, or by right-clicking and selecting Extract from the pop-up menu.

This book is here to help you get your job done. In general, you may use the code in this book in your programs and documentation. You do not need to contact us for permission unless you're reproducing a significant portion of the code. For example, writing a program that uses several chunks of code from this book does not require permission. Selling or distributing a CD-ROM of examples from Make: books does require permission. Answering a question by citing this book and quoting example code does not require permission. Incorporating a significant amount of example code from this book into your product's documentation does require permission.

We appreciate, but do not require, attribution. An attribution usually includes the title, author, publisher, and ISBN. For example: "*Getting Started With Sensors* by Kimmo Karvinen and Tero Karvinen (Maker Media). Copyright 2014, 978-1-4493-6708-4."

If you feel your use of code examples falls outside fair use or the permission given here, feel free to contact us at *bookpermissions@makermedia.com*.

Safari® Books Online

Safari Books Online is an on-demand digital library that delivers expert content in both book and video form from the world's leading authors in technology and business.

Technology professionals, software developers, web designers, and business and creative professionals use Safari Books Online as their primary resource for research, problem solving, learning, and certification training.

Safari Books Online offers a range of product mixes and pricing programs for organizations, government agencies, and individuals. Subscribers have access to thousands of books, training videos, and prepublication manuscripts in one fully searchable database from publishers like Maker Media, O'Reilly Media, Prentice Hall Professional, Addison-Wesley Professional, Microsoft Press, Sams, Que, Peachpit Press, Focal Press, Cisco Press, John Wiley & Sons, Syngress, Morgan Kaufmann, IBM Redbooks, Packt, Adobe Press, FT Press, Apress, Manning, New Riders, McGraw-Hill, Jones & Bartlett, Course Technology, and dozens more. For more information about Safari Books Online, please visit us online.

How to Contact Us

Please address comments and questions concerning this book to the publisher:

Make:
1005 Gravenstein Highway North
Sebastopol, CA 95472
800-998-9938 (in the United States or Canada)
707-829-0515 (international or local)
707-829-0104 (fax)

Make: unites, inspires, informs, and entertains a growing community of resourceful people who undertake amazing projects in their backyards, basements, and garages. Make: celebrates your right to tweak, hack, and bend any technology to your will. The Make: audience continues to be a growing culture and community that believes in bettering ourselves, our environment, our educational system—our entire world. This is much more than an audience, it's a worldwide movement that Make: is leading—we call it the Maker Movement.

For more information about Make:, visit us online:

Make: magazine: *http://makezine.com/magazine/*
Maker Faire: *http://makerfaire.com*
Makezine.com: *http://makezine.com*
Maker Shed: *http://makershed.com/*

We have a web page for this book, where we list errata, examples, and any additional information. You can access this page at: *http://bit.ly/get-start-sensors*.

To comment or ask technical questions about this book, send email to: *bookquestions@oreilly.com*

Acknowledgments

The authors would like to thank Hipsu, Marianna, Nina, and Valtteri.

1/Sensors

Sensors surround you in daily life. The world is full of them: from passive infrared sensors in motion detectors, to CO_2 detectors in air conditioning systems, and even tiny accelerometers, GPS modules, and cameras inside your smartphone and tablet—sensors are everywhere! The variety of sensor applications is remarkable.

It's safe to assume that if an electronic device is considered "smart," it's full of sensors (Figure 1-1). In fact, thanks to the proliferation of smart devices, especially phones, the price of sensors has been driven to affordability. Not only is it economically viable to add advanced sensors to your projects, but they vastly expand the kinds of projects you can make.

You'll learn about sensors in this book by making small projects and reflecting on the experience. It's more fun to build first and discuss later, but both are equally important. It's best to avoid the temptation to only build projects and skip the conceptual sections.

Getting started with sensors is easy, and only the sky is the limit. Electronics challenge some of the best brains daily and produce new innovations and dissertations. On the other hand, even a child can get started with some guidance.

If you don't know much about sensors yet, try to remember what it feels like now. After you've tackled some challenges and built a couple of gadgets, many dark mysteries of sensors will probably seem like common sense to you.

This book is suitable for anyone with an interest in sensors (see Figure 1-2). After you've built the gadgets and have read this book, you can get ideas for bigger projects from our book *Make: Arduino Bots and Gadgets* or learn more advanced sensors in *Make: Sensors*. For a wider view of the basics, see *Getting Started with Arduino, 2nd Edition* by Massimo Banzi, *Getting Started with Raspberry Pi* by Matt Richardson and Shawn Wallace, or *Make: Electronics* by Charles Platt.

Figure 1-1. *Various sensors: infrared proximity, rotation, brightness, button, temperature, and distance*

What are sensors? Sensors are electrical components that function as input devices. Not all inputs are explicitly sensors, but almost all inputs use sensors! Consider your computer mouse or trackpad, a keyboard, or even a webcam; these are not sensors, but they definitely use sensors in their design. More abstractly, you can frame sensors as a component to measure a stimulus that is external to the system it is in (its environment). The output data is based on the measurement. For example, when you type at a keyboard, the letter that appears on your screen (the output) is based on the measurement (which switch, or key, you pressed on the keyboard). How many letters appear on screen is based on another measurement (how long you keep the key pressed).

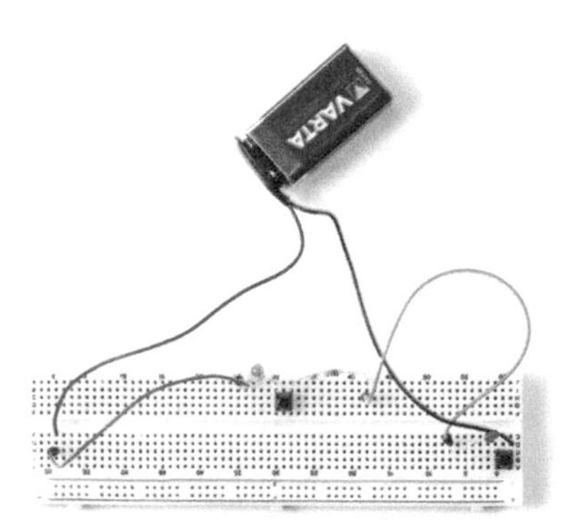

Figure 1-2. *Simple AND connection with buttons, built and designed by a four-year-old with help from an adult*

The first project uses a photoresistor to measure light. Without the photoresistor (or similar sensor), there is no way the circuit can know how bright the light is in the environment. By adding the sensor, your circuit knows something it didn't know before.

All of the projects in this book *evaluate* a particular stimulus within the environment. None of this would be possible without sensors. Let's get building so you can experience the inputs and outputs that sensors provide to projects.

Project 1: Photoresistor to Measure Light

Light in an environment is quite informative: you can determine what time of day it is based on the sun's angle, you operate a car more safely at night when its lights are on, and people who do not experience enough light in daily life can become depressed with seasonal affective disorder. As such, light influences many aspects of your life and it's fun to measure it, too.

The simplest sensor for detecting light is a *photoresistor*. It's not uncommon to also encounter another name for the exact same sensor: *light-dependent resistor* (LDR). The component works by changing its resistance based on the amount of light hitting it.

Now that you know the right sensor to use, the next question to think about is how to process the sensor's measurements. If you've ever worked with a light-emitting diode (LED) , shown in Figure 1-3, you might know that resistance is an electrically important consideration. For example, if you've ever used a larger-value resistor for the LED than a project called for, you've seen that too much resistance can restrict an LED from illuminating. This same basic observation is applicable to this project.

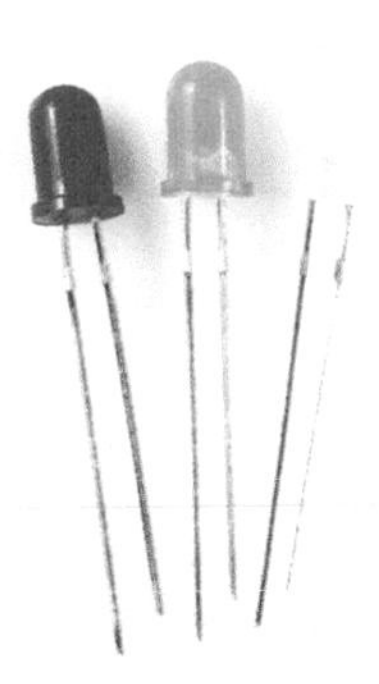

Figure 1-3. *LEDs*

The circuit is designed so that an LED is dependent on the photoresistor's measurement. Too much resistance and the LED simply will not turn on. Enough discussion—it's time to build! Figure 1-4 shows the finished project.

Figure 1-4. *The finished photoresistor project*

Parts

You need the following parts for this project:

- Photoresistor
- 5 mm red LED (different LEDs will work differently with this circuit; later, you'll learn a more sophisticated way to fade LEDs)
- 470 Ω resistor (four-band resistor: yellow-violet-brown; five-band resistor: yellow-violet-black-black; the last band will vary depending on the resistor's tolerance)
- Breadboard
- 9 V battery clip
- 9 V battery

All of these parts, except the 9 V battery and 470 Ω resistor, are available in the Maker Shed Mintronics: Survival Pack (*http://bit.ly/mintron-sp*), part number MSTIN2. You can use two of the 220 Ω resistors in series or one 1 kΩ resistor in place of the 470 Ω resistor; both of these are available from electronics retailers such as RadioShack.

Build It

Here are the steps for building this project:

1. Orient your breadboard so that it is wider than it is tall, as shown in Figure 1-5.

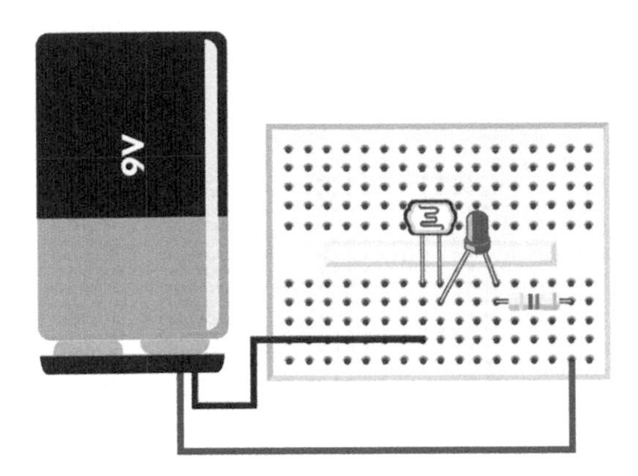

Figure 1-5. *Circuit digram for photoresistor project*

2. Look at your LED and determine which lead has a flat side above it on the colored plastic housing—this indicates the negative lead of the LED (the negative lead is also the shorter of the two), as shown in Figure 1-6. LEDs have a certain *polarity* and putting them in backward might damage them.

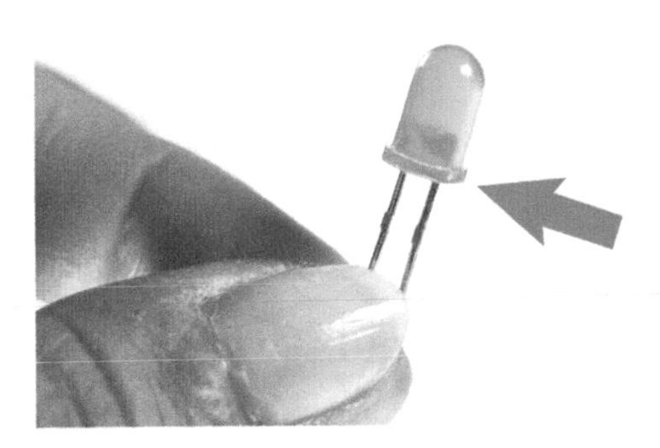

Figure 1-6. *Negative leg of the LED*

3. Insert the photoresistor so that the negative lead of the LED and one of the photoresistor leads occupy the same column. The second (positive)

LED lead should occupy its own column for now. Refer back to Figure 1-5 to see how they should be arranged.

Do you see the gap in the middle of the breadboard in Figure 1-7? That gap separates the two groups of columns, and there's no connection across it. If you want two leads in the same column to be connected, be sure they are on the same side of the gap.

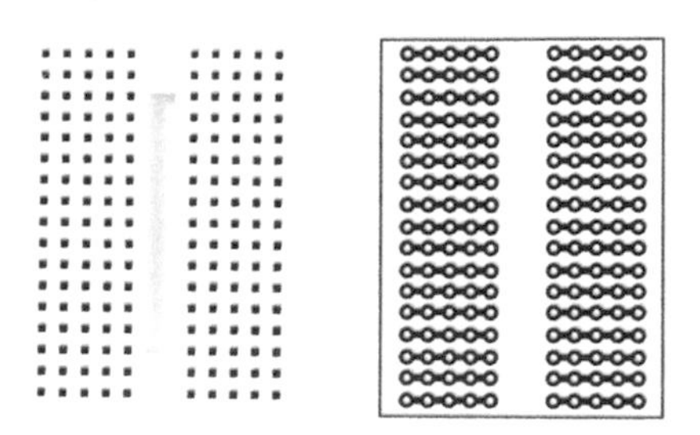

Figure 1-7. *Breadboard layout*

4. Connect the 470 Ω resistor to the column with the positive LED lead and make sure it's not the same column that already has the photoresistor and LED's negative lead in it. Make sure that the resistor's other lead is in a separate column.
5. Attach the black wire from the battery clip to the column that contains only a photoresistor lead.
6. Insert the red wire from the battery clip to the column that contains only a resistor lead.
7. Double-check the steps and if everything looks like Figure 1-5, connect the 9 V battery.

That's it. You've built your first sensor circuit. Congratulations!

Discussion: Photoresistors

It might not seem like the circuit is doing much. That's because the light levels probably have not changed much in your room. Put your finger over the photoresistor and watch the LED closely. Did anything seem to change with the LED? There should have been some change in the LED brightness. Try adding more light to the photoresistor. The opposite happened, right? Now that you've seen the photoresistor in action, how would you describe what is going on when you expose the photoresistor to more light in terms of its resistance? Is the resistance increasing or decreasing when the sensor is exposed to more light?

Here's what's going on in the circuit. The more light hits the photoresistor, the lower its resistance. If the room is quite bright, then the LED will be quite bright. If light is low in the room, the sensor resists the current flow, which is expressed by the LED getting dimmer.

The current flow through the sensor controls how bright the LED will shine. That's because the circuit is wired so that all current to the LED must pass through the photoresistor first.

The photoresistor, as you just learned, is a *resistive sensor*. There are many types of resistive sensors; this category of sensors is used to measure much more than luminosity. As you continue reading and you encounter new sensors, it's a good idea to think how the stimuli are measured and especially how the output is structured. None of the sensors will output data that is in a convenient format for end-user consumption. Instead, you will need to decide how to express or format the raw sensor data output in a way that makes sense to users.

Another sensor type is *electro-mechanical*. These sensors do not manifest changes in voltage or current, but rather by a change in their physical properties. The thermostat in your home or apartment is a great example (unless you have a digital thermostat). When the room temperature changes, a thermostat's *bimetallic coil* will expand or contract depending on whether the room's temperature increases or decreases. The sensor is actually physically expressing itself by changing shape! But even these sensors may trigger an electronic sensor (for example, a thermostat's bimetallic coil is usually connected to a tilt switch that turns the heat on or off).

Interactive Sensor Control

"Project 1: Photoresistor to Measure Light" on page 3 used a sensor in a way that didn't directly involve interaction with a human. Sure, you were the one who changed the lighting in the room to force the photoresistor to change its resistance, but it could have just as easily been the setting or rising of the sun. There are many sensors that you'll manipulate directly; you'll see these in later chapters.

Going Forward

In all the projects in this book, you'll be building small systems that collect input data by taking measurements with a sensor. The systems will do something that processes that input data, and then take action (the output of the system). At first, you'll just build things with electronic components, but later in this book, you'll use Arduino and Raspberry Pi to handle the processing.

When you use Arduino and Raspberry Pi, you'll write code that does a lot of the work for you. The benefit will become very clear because you'll find that you can change the way you respond to an input without having to rewire your circuit.

Suppose you want to challenge your friend to see who can press a force-sensing resistor down more firmly. All you would need to do is add some extra lines of code to the Arduino sketch and send the new code. How would you accomplish the same trickery if you didn't have the option to alter code? It certainly would be more challenging, and at the very least you would have to move a few things on the breadboard. And how would you display the score for the contest results? It turns out that an Arduino handles quite a lot of work for you!

But this isn't the whole story of sensors. We don't want you to think that as sensor systems increase in physical complexity, their programs become more complicated. Rather, we want you to think of this as an issue of applicability: what's the best design to accomplish your goal?

2/Basic Sensors

Get ready to wire a few more sensors and learn a bit more theory. The circuits you will build in this chapter do not need to be programmed, which is to say, they will not use an Arduino or a Raspberry Pi. The sensor's data will be "interpreted" by the electrical properties of each component in the circuit design.

A sensor is a physical input to a circuit. Sensors are *transducers*. The process of converting sensed energy into another form is called *transduction*. For example, a light sensor transduces luminosity into resistance.

Another type of transducer is an *actuator*. Rather than reacting to something in the environment, an actuator makes something happen. An actuator is a physical output of a circuit. An LED, for example, transduces electrical current into light; a speaker turns it into sound.

You'll use both types of transducers (sensors and actuators) in the projects in this book. So if someone asks you how a sensor exerts control over a circuit's simply reply, "Transduction!"

Project 2: A Simple Switch

A switch can connect and disconnect the flow of electricity. It's just like a button, except that it stays where you leave it. If you switch it off, it stays off. If you switch it on, it stays on.

Parts

You need the following parts for this project:

- A switch
- A wire with alligator clips
- Two 1.5 V batteries
- Battery holder with wire leads
- 5 mm red LED

- 470 Ω resistor (four-band resistor: yellow-violet-brown; five-band resistor: yellow-violet-black-black; the last band will vary depending on the resistor's tolerance)
- Breadboard

Some toggle switches have three leads. These *single pole double throw* (SPDT) switches are used in cases where you want the switch to choose between two different paths for electricity to flow. If you only wire up two of the leads, you're just using it as a regular single pole single throw (SPST) switch.

SPDT switches are used when a simple on-off switch won't do the trick, such as the pair of light switches commonly found at the top and bottom of a staircase. Because of a multiway switching wiring scheme, flipping either switch will toggle the state of the light.

Build It

Figure 2-1 shows the circuit diagram for this project. Here's how to wire it up:

1. Orient your breadboard so that it is taller than it is wide, as shown in Figure 2-1.

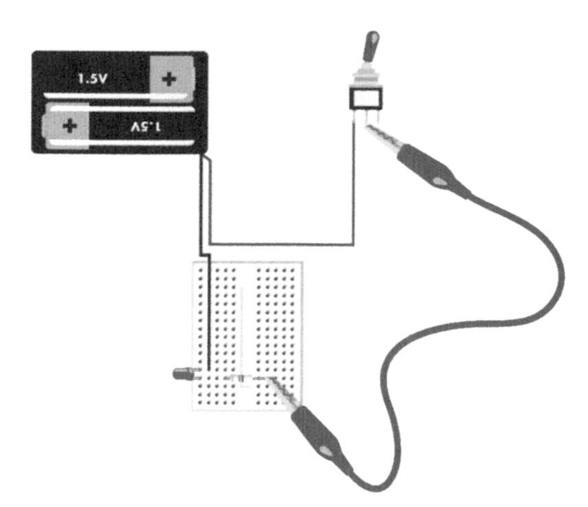

Figure 2-1. *Circuit diagram for the simple switch circuit; the short lead connects to the black wire*

2. Insert the red LED into the breadboard and keep track of which side is positive (the longer lead) and which is negative (the shorter lead).

3. Insert one lead of the 470 Ω resistor into the row containing the LED's positive lead.

Keep in mind that the gap in the middle of the breadboard separates the two groups of rows, and there's no connection across that gap. If you want two leads in the same row to be connected, be sure they are on the same side of the gap.

4. Connect the alligator wire from the other resistor lead to the middle lead of the switch.
5. Connect the red positive wire from the battery holder to either side lead of the switch using the following technique (see Figure 2-2):
   a. Push the wire through the hole.
   b. Fold the wire into itself.
   c. Twist the folded wire to secure it.

Figure 2-2. *Tying a wire to the switch leads*

6. Insert the batteries into the holder.

With everything wired up, test the switch and see if your LED turns on and off as you'd expect from a switch.

Now you can add a power switch to any of your projects. No more removing the battery when you are done playing with a prototype.

Troubleshooting

If this project doesn't work, try checking the following things:

- If your switch has three leads, did you connect the switch's center and one side lead (rather than using two side leads)?
- Make sure that the alligator clip only touches the middle lead of the switch. If the alligator clip short-circuits two leads, the LED might be lit constantly, even if you flip the switch.
- Is the battery wire connected securely? If you find the LED is never lit, try pressing on the battery wire connection and flipping the switch. For a permanent fix, twist the switch more to make the wire tighter.

An LED Needs a Resistor

All the circuits in this book work with a normal red LED and a 470 Ω resistor. In a pinch, any resistor with a brown multiplier band (third stripe) will do.

What if you have a really fancy LED, such as a blue one or a superbright LED? If you have a general idea what the LED should look like when lit, you can just try out resistors. Start with the common 470 Ω. If the LED is too dim, pick a weaker resistance.

Most LEDs don't break easily, but they have very low resistance. So if you forget the current limiting resistor, the LED might drain your battery and cause some components to overheat and break. Even if you use an online LED resistor calculator (there are many out there; just Google it or look in your phone's app store) and it gives you a zero (or less!), you should still use a very small resistor.

Third Band Trick

We use a shortcut called the *third band trick* to pick out resistors. Hold a resistor so the tolerance band is on your right—these are usually gold or silver bands. The third band in from the left is known as the multiplier band.

Some resistors have five bands, not four, although they are uncommon. But all the same, the multiplier band is the second-to-last band. For a five-band resistor, it's the fourth band.

Our rule of thumb (when working with 3 V to 5 V) for choosing a resistor by its third band: pick brown if you are protecting an LED, green if you need a pull-up resistor.

Project 3: Buzzer Volume Control

In many projects, you need the ability to slightly adjust input or output rather than toggling between on and off. A *potentiometer* will allow you to operate between the extremes of on and off, just as the photoresistor did in "Project

1: Photoresistor to Measure Light" on page 3. Potentiometers, also known as *pots*, are often used as volume controls for audio devices.

Parts

You need the following parts for this project, which are available in the Maker Shed Mintronics: Survival Pack (*http://bit.ly/mintron-sp*), part number MSTIN2 (you will need to purchase the 9 V battery and piezo buzzer separately:

- DC piezo buzzer (Maker Shed part number MSPT01 (*http://bit.ly/piezo-buzz*))
- 10 K potentiometer
- 9 V battery
- Battery clip
- Breadboard

You are using a 9 V battery in this project because common piezos typically operate with a range of 6 V to 18 V or more. Because we are going to be increasing and decreasing the current with a potentiometer, using a 9 V source voltage gives us room to increase or decrease the voltage and still produce audible output.

Build It

Here are the steps for building this project:

1. Orient your breadboard so that it is wider than it is tall, as shown in Figure 2-3.
2. Connect the negative wire of the piezo buzzer to any free column of holes on the breadboard and then insert the black (negative) battery clip wire into the same column.
3. Insert the potentiometer into the breadboard, but make sure you don't put it in the same column as the negative wires you just connected. Put the battery clip's red (positive) lead into the same column as the potentiometer's middle lead.

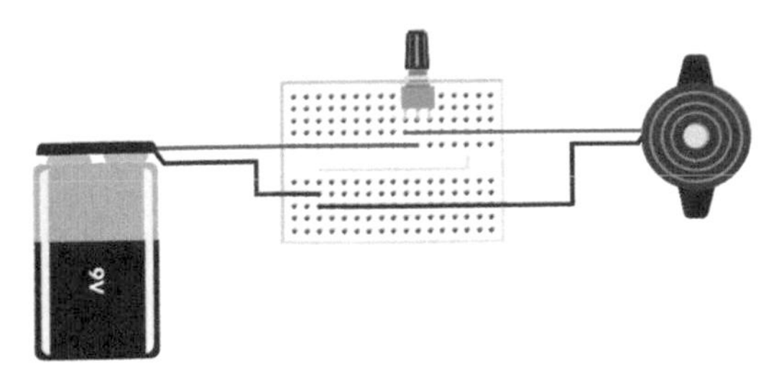

Figure 2-3. *Circuit diagram for buzzer volume control*

4. Next, connect the piezo's positive lead to one of the potentiometer's outer pins. It doesn't matter which outer pin you choose, just don't pick the middle one—that's not an outer pin.
5. Clip in the 9 V battery and turn the potentiometer left to right to change the output level of the piezo buzzer.

Now you can add loud noise to any of your projects. You also know how to use a potentiometer as an adjustable resistor. You will soon learn about many other sensors that report their value with resistance, just as a potentiometer does.

Troubleshooting

If this project doesn't work, try checking the following things:

- Does your piezo buzzer work? Connect it directly to the battery to make sure it works (the piezo black lead goes to the battery's black lead, and red goes to red).
- Did you connect the pot correctly? We're using the pot as a simple adjustable resistor, so we must connect one wire to the center lead and the other to either of the side leads.
- Check connections one by one. Start from battery positive or negative and go through the whole circuit.

Project 4: Hall Effect

A hall switch senses change in the magnetic field and detects if there is a magnet nearby. A very common use for this is a bike speedometer that identifies how often a rapidly rotating object passes by the sensor. Another typical application is a door burglar alarm like the simplified version we're going to build next.

Figure 2-4 shows the NJK-5002A Hall effect switch, which you can find from sellers on eBay (*http://ebay.com*) or AliExpress (*http://aliexpress.com*) for under $10. If you buy the similar NJK-5002C, you will have to change the way you wire the circuit (described later).

Figure 2-4. *Hall effect switch with magnet*

Parts

You need the following parts for this project:

- DC piezo buzzer
- 9 V battery
- Battery clip
- Breadboard
- NJK-5002A Hall effect switch

Build It

The NJK-5002A acts as a switch, and has connections for positive voltage (brown), negative (blue, although most components use black for negative), and a connection that outputs positive voltage (black, despite the fact that black usually indicates negative) when a magnet is held to the sensor.

If you are using the NJK-5002C, black will go negative rather than positive when a magnet is held to the sensor.

Here are the steps for building this project:

1. Orient your breadboard so that it is wider than it is tall, as shown in Figure 2-5.

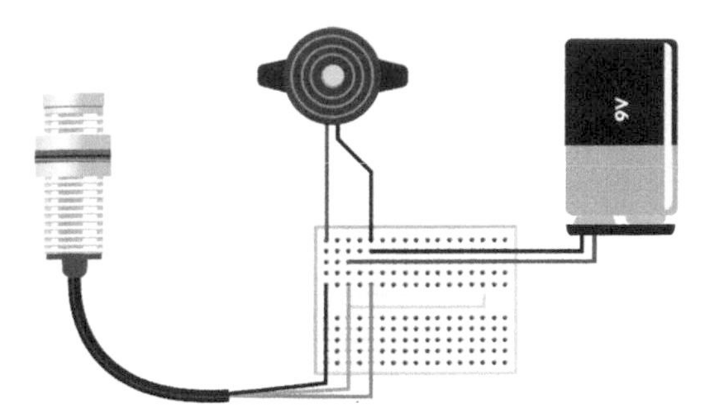

Figure 2-5. *Circuit diagram for the Hall sensor alarm*

2. Insert the black (output) Hall effect switch wire into any free column of holes on the breadboard and then insert the red positive wire of the piezo buzzer into the same column.

   If you are using the NJK-5002C instead of the NJK-5002A, connect the black negative wire from the piezo here instead of the red positive wire.
3. Insert the red positive wire of the 9 V battery clip into any free column of holes on the breadboard and then insert the brown (positive) Hall effect switch wire into the same column.

   If you are using the NJK-5002C instead of the NJK-5002A, insert the red positive wire from the piezo into this column as well.
4. Insert the blue (negative) Hall effect switch wire into any free column of holes on the breadboard and then insert the black negative wire from the battery clip into the same column.

   Insert the black negative wire from the piezo into this column. (If you are using the NJK-5002C, you don't need to make another connection to it in this step.)
5. Clip in the 9 V battery and hold a magnet to the switch to make the buzzer sound.

Now that you've learned how to work with a Hall effect switch, you can use it in your projects in many ways. You can use it to detect when two things are brought together or separated. For example, you could put a magnet in a door, and a Hall effect switch in the doorframe to know when it is opened or closed.

It can also be used to measure rotational speed: if you embed a magnet in a stationary bicycle wheel, you could use a Hall effect switch to measure how fast you are pedaling. You could use the speed as input to a video game to make your exercise sessions more interesting.

Troubleshooting

If this project doesn't work, try checking the following things:

- Try flipping the magnet over if it doesn't work. The switch won't be triggered unless you press the right pole of the magnet against it.
- Did the LED that's built into the sensor light up? If not, it's probably not wired right.
- Are you using the correct model of Hall switch? The build instructions tell you how to connect two popular models (NJK-5002A and NJK-5002C). If you're using a different model, check the datasheet for information that will help you connect it correctly.
- Make sure the piezo buzzer is working by connecting it directly to the battery. Does it make a sound?

Project 5: Firefly

Night falls, and fireflies start to glow, slowly fading in and out. In case your garden doesn't have any real fireflies, you can build your own. You'll use a light-dependent resistor to detect darkness, and a 555 timer circuit to continuously fade an LED in and out.

It's time to venture beyond basic components and leap to integrated circuits (ICs). You'll still use the components from the previous project, but also learn to use one of the most popular ICs, the 555 timer.

First, you'll make an LED light up when it's bright outside. Then you reverse this, making your LED light up in darkness. You'll make the LED fade in and out with the 555. Finally, you'll combine the circuits to make the firefly. It turns on the LED fading in darkness.

By building the 555 firefly in distinct, testable steps, you will learn a systematic approach to building analog circuits. You won't learn analog circuit design here, but at the end of this project you'll get some ideas for simple modifications. Designing completely new gadgets is easier with Arduino, which you'll see later in the book.

Integrated Circuits

ICs vary in complexity and price. The 555 timer we use here is a cheap, comparatively simple IC: it only has about 40 components inside, most of these being transistors and resistors. Still, imagine if you didn't have the luxury of using a 555 timer. Your projects would balloon in parts and the wiring would not be an easy process. If you think 40 components is a lot, you will be taken aback to learn that there are billions of components in ICs such as a processor or memory chip. The most complex IC we will be using in our sensor projects is the Broadcom BCM2835 *system-on-chip* found in the Raspberry Pi.

This book teaches you the things you need to know to use the components in this book. If you want to know more, you can search the Web for the name of the IC and the word "datasheet" (e.g., "CA555E datasheet"). Datasheets are sometimes tedious to read, but they are the authoritative source on how the IC works.

Pin numbering can be found on the datasheet. Usually, pins are numbered counter-clockwise from the notch. Hold the IC, pins down, so that you look at it from the top. Locate the polarity mark: a half-moon notch, a corner triangle, or a dot on pin 1. Pins are numbered counter-clockwise from that mark, starting from pin 1.

For the CA555E, hold the half-moon notch to the left and the pins away from you. See Figure 2-6. Pin 1 is the bottom left, and the numbers increase counter-clockwise. Thus, the last pin (8) is at top left. To make your life easier, the circuit diagrams in this book have small pin numbers on the 555 chips.

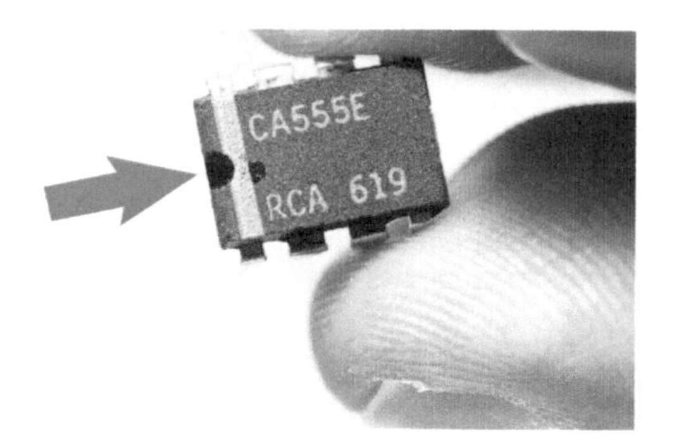

Figure 2-6. *Pins are numbered counter-clockwise from the notch*

555 Timer IC

The 555 timer IC (shown in Figure 2-7) is a multipurpose chip. You can configure it as a timed delay, an oscillator, or as a flip-flop.

Modes of 555 operation are:

- Leave the output on for a while (*monostable*).
- Toggle the output (*bistable*, also known as flip-flop).
- Blink the output without any input (*astable*).

Figure 2-7. *555 timer integrated circuit*

Configuring a 555 timer is *not* analogous to the programming you will do on an Arduino or Raspberry Pi, where lines of code are compiled and executed. Instead, the 555 timer configuration is done by connecting certain components such as resistors and capacitors to the 555. Hardware sets up the 555, not software.

Building and configuring gadgets by coding with Arduino is easier than the component-by-component approach you use with the 555. Here, we have already designed the circuit for you and made the necessary calculations, making it a simple task to get familiar with the 555.

Light Up an LED When It's Bright

A light-dependent resistor (LDR) cuts off electricity when it's dark. When it's well lit, the resistance is low and current passes through. The brighter it is, the more current goes through.

How do you light up an LED when it's bright? Just connect an LDR in series with the LED. The LDR works just like any analog resistor, so you could use a potentiometer in its place.

Sound familiar? That's because you've built this circuit earlier (see Figure 2-8). Refer to the building instructions in "Project 1: Photoresistor to Measure Light" on page 3.

Put your finger over the LDR, and the LED goes dark. Aim a bright light at the LDR, and the LED is brightly lit.

Who needs light in a bright room? Can't you make it light up in the darkness?

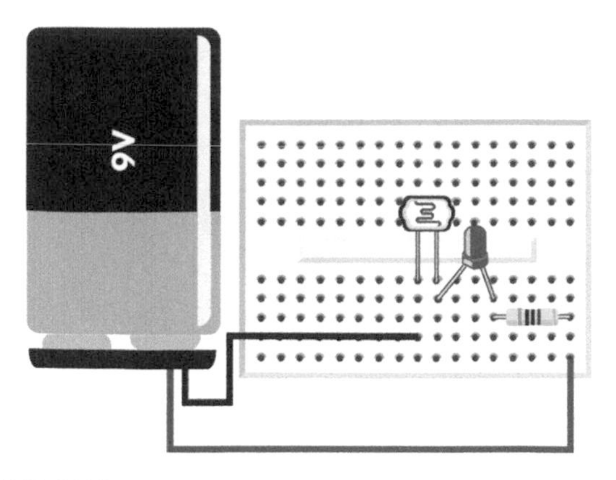

Figure 2-8. *LED/LDR circuit*

Jumper Wires

Jumper wires, also known as hookup wires, allow you to connect components to each other on a breadboard (Figure 2-9).

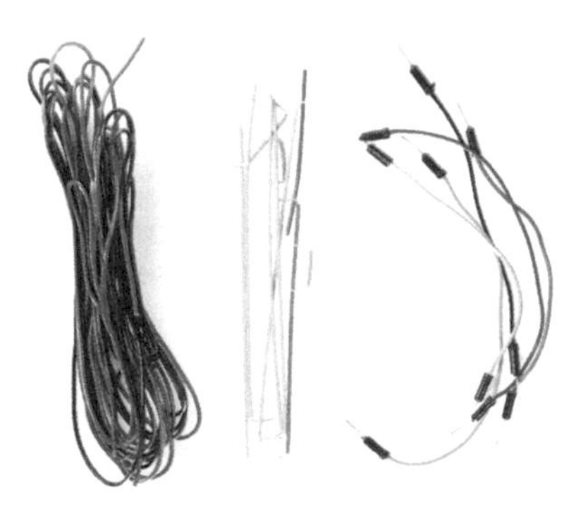

Figure 2-9. *Jumper wires*

These wires are not fancy and you can even make them yourself by stripping the ends off of insulated solid core thin wire. Just because you can make jumper wires doesn't mean it is not worth buying a pre-made collection of jumper wires. They are sold in various colors, and their ends are straight and reinforced so it's easy to push them into the breadboard.

Jumper wires are available in a number of colors, lengths, and connector housings. The core of the wire refers to the part surrounded by plastic insulation and can be stranded or solid. You want solid core wire because the breadboard clips need to latch onto something solid.

In an attempt to make life easier, jumper wires come in a variety of colors. There is nothing electrically different about these hued wires, but typically black is used for wires connected directly to a ground terminal (0 V, GND) and red is used for direct connections to a positive terminal (+5 V). We strongly recommend you reserve black for negative/ground and red for positive, but feel free to use all the other colors any way you want.

Light to Darkness

You can use a transistor and a resistor to invert the effect of the LDR.

A transistor is an amplifier (see "Transistors" on page 22). In this circuit, you use the most common transistor, a bipolar NPN transistor. The amplifier circuit here is the common emitter amplifier, the most common transistor amplifier (see "Common Emitter Amplifier" on page 23).

Controlling base current to the transistor is done with two resistors. One of the resistors is connected to the plus terminal of the battery, and pulls the voltage up. The other resistor, the LDR, is connected to ground (minus). When it's bright, LDR has very low resistance. Thus, when it's bright, the base of the transistor gets pulled to ground, so it gets no current: zero amplified is still zero and the LED stays dark.

The connection using the two resistors is called a *voltage divider*.

Parts

You need the following parts for this project:

- BC547 transistor
- Light-dependent resistor (LDR)
- 470 Ω resistor (four-band resistor: yellow-violet-brown; five-band resistor: yellow-violet-black-black; the last band will vary depending on the resistor's tolerance)
- 5 mm red LED
- 9 V battery
- 9 V battery clip
- Breadboard

All of these parts, except the 9 V battery, 470 Ω resistor, and BC547, are available in the Maker Shed Mintronics: Survival Pack (*http://bit.ly/mintron-sp*), part number MSTIN2, from . You can use two of the 220 Ω resistors in series or one 1 kΩ resistor in place of the 470 Ω resistor; both of these are available from electronics retailers such as RadioShack.

Build It

Build the circuit as shown in Figure 2-10. Covering the LDR with your finger lights up the LED. Point a flashlight at the LDR and see the LED go dark.

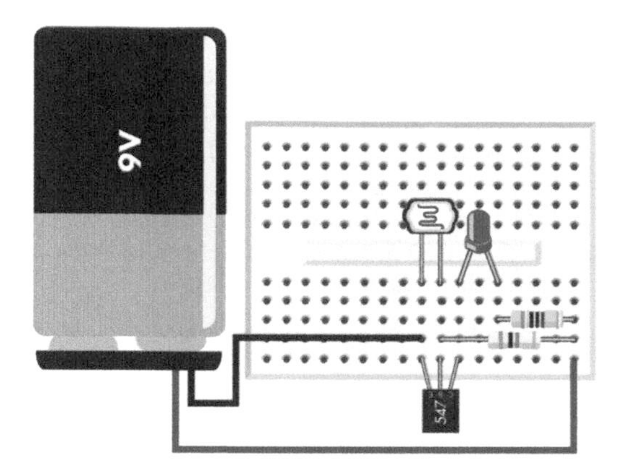

Figure 2-10. *LDR inverter circuit*

Transistors

A transistor (see Figure 2-11) is an amplifier, taking small signals and making them larger. It can also function as a switch, because a small power can control a larger power.

Transistors are actually the basis of all digital electronics. Your computer's CPU, the ATMega microcontroller in an Arduino board, and the *system-on-a-chip* on the Raspberry Pi have millions, even hundreds of millions of transistors arranged in patterns to create digital logic.

Microcontrollers, such as Arduino and Raspberry Pi, use very little power. If you want to control something that needs more power, such as a big motor, a transistor is one way to do that.

Figure 2-11. *A transistor*

In most prototyping projects, you don't need an in-depth understanding of transistor circuits. For example, you can use an Arduino and a servo (easy) instead of trying to control a motor with transistors (harder). That's why it's enough to get the high-level understanding of a transistor as an amplifier. You won't need to be able to design any transistor circuits after reading this short explanation about transistors.

The most common type of transistor is the NPN-type bipolar junction transistor (BJT). We'll only talk about these common NPN BJT transistors here, as it's much easier to learn the others once you know the NPN BJT transistor. The BC547 transistor you use in this chapter is a typical NPN BJT transistor.

An NPN BJT transistor has three leads: emitter E, base B, and collector C.

Common Emitter Amplifier

Emitter E is the common ground, the negative. The arrow points to the negative lead. As it says in the NPN mnemonic, "Not Pointing iN," the arrow in the NPN transistor symbol is pointing out to negative. The arrow is simply pointing the direction of current from positive to negative (see Figure 2-12).

The small controlling base current (BE) flows from base B to emitter E. This base current is the weak signal to be amplified.

The large collector current (CE) flows from collector C to emitter E. This collector current only flows if there is base current.

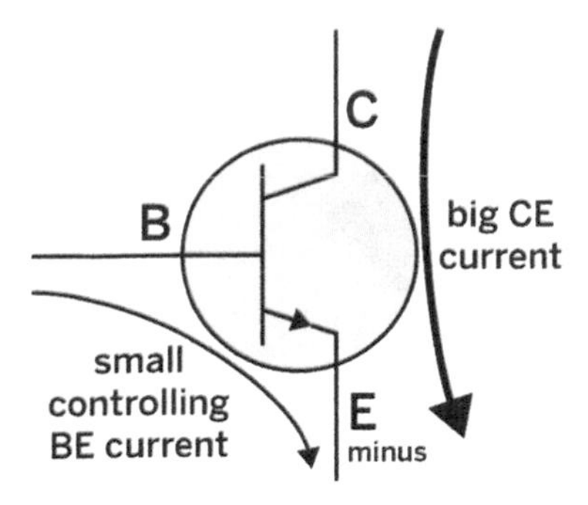

Figure 2-12. *NPN BJT transistor*

To give you an example of a transistor in action, consider a transistor where the base is at 0 V. There is no base-emitter current flowing, so the big collector-emitter current is blocked. When you apply current to the base, a small base-emitter current starts flowing. This allows the big collector-emitter current through, and could start a big motor or a bright light.

Take the transistor from this project, BC547, with leads down and the flat text side facing you. The collector is on the left, base is center, and the emitter is on the right.

Fading an LED

What could be easier than fading a light? Actually, fading an LED is tricky business. If you gradually add voltage, first it's off and then it's on. For a dimly lit LED, there is only a small voltage range available.

With microcontrollers, such as an Arduino, dimming is done by blinking the LED faster than the eye can see. Here, with analog electronics, you can use a transistor to help you control the small part of a voltage range where the LED is between on and off (see Figure 2-13).

In this circuit, the potentiometer sets the voltage of the transistor's base (B). The transistor is used in the common collector amplifier, where the small BC current controls the bigger EC current.

Build the circuit first, and test how you can use most of the potentiometer's range to control the brightness of the LED.

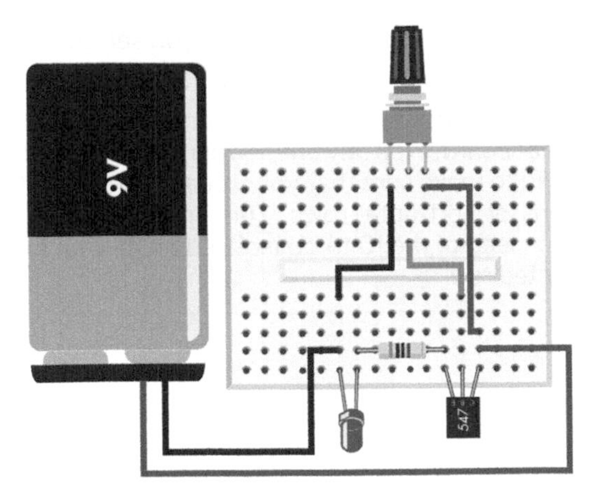

Figure 2-13. *Fading an LED with an NPN transistor*

Is there some doubt in your heart that the transistor is really needed? You can easily test it by building a worse, alternative setup (see Figure 2-14). Put a potentiometer in series with the LED. When you try out that setup, you'll see that only a small part of the potentiometer range dims the LED. How well (or poorly) the potentiometer works without the transistor depends on the LED you're using.

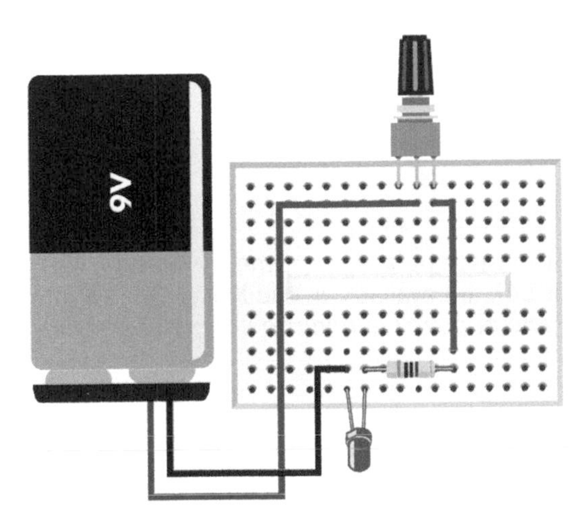

Figure 2-14. *Bad way of trying to fade an LED*

555 Fading Blink

Your firefly should continuously fade between lit and dark. You can use the 555 timer to fade the LED.

Build the circuit, then enjoy the explanation as you see your LED fading.

The previous projects can be accomplished on any type of breadboard, even tiny ones. When you use an IC, however, you need to make sure that your breadboard is split into two sections of horizontal rows. This breadboard layout allows for each lead of the IC to be isolated on its own electrically connected row.

Parts

You need the following parts for this project:

- BC547 transistor
- 470 Ω resistor (four-band resistor: yellow-violet-brown; five-band resistor: yellow-violet-black-black; the last band will vary depending on the resistor's tolerance)
- 5 mm green LED
- 9 V battery
- 9 V battery clip
- Breadboard
- CA555E, NE555, or compatible 555 timer integrated circuit
- 10 kΩ potentiometer
- 100 uF capacitor
- Jumper wires

All of these parts, except the 9 V battery, 470 Ω resistor, and BC547, are available in the Maker Shed Mintronics: Survival Pack (*http://bit.ly/mintron-sp*), part number MSTIN2.

Build It

Build the circuit as shown in Figure 2-15.

In this circuit, the 555 timer is used in astable mode. The 555 output fades between 3 V and 6 V. Adjusted with a transistor amplifier, this makes the LED fade in and out.

The input pins of the 555 are 6 (THRES) and 2 (TRIG). The two pins are connected together. Pin 2 (TRIG) detects a voltage drop below 3 V and turns pin 3 (OUT) to 7 V, so that output voltage starts rising. Pin 3 (OUT) is connected to the capacitor through the potentiometer, so the 7 V voltage starts loading the capacitor. You can turn the potentiometer to choose how fast the capacitor charges.

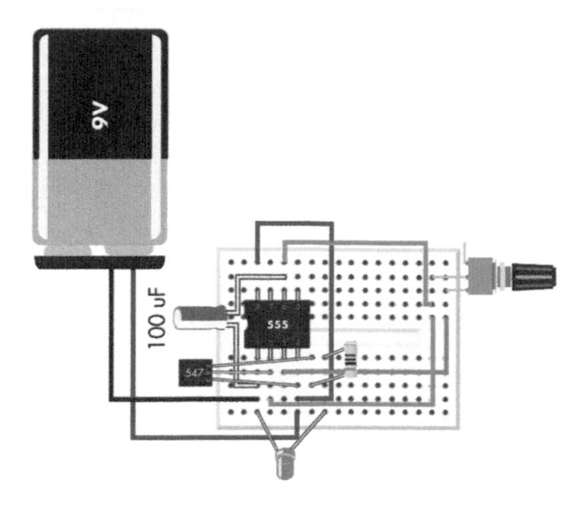

Figure 2-15. *555 fader circuit*

When the voltage on pin 6 (THRES) rises to 6 V, the 555 detects this and turns OUT low (to GND), so the voltage starts falling down. As you remember, the capacitor is connected to pin 3 (OUT) through the potentiometer, so the potentiometer chooses how fast the capacitor discharges.

As soon as the capacitor has discharged enough to bring the voltage to 3 V, pin 2 (TRIG) pin toggles pin 3 (OUT) to 7 V. The capacitor starts to charge, and the cycle starts over.

This way, the output of the 555 circuit keeps oscillating between 3 V and 6 V. Because it takes time to charge the capacitor, your circuit produces a triangle wave.

You probably want to connect an LED to the output of your new 555 oscillator. Or even better, connect the fading LED circuit with a transistor (see "Fading an LED" on page 24). So where do the two wires go? See Table 2-1. The output of the astable 555 circuit are GND (the 0V minus wire) and the point that connects to the capacitor, pin 2 (TRIG), and pin 6 (THRES).

Table 2-1. *555 pins used in this project*

Number	Name	Purpose
1	GND	Ground, 0 V, goes to the minus terminal of the battery
2	TRIG	If pin 2 (TRIG) goes below 3 V, turns pin 3 (OUT) to 7 V
3	OUT	Either 0 V or 7 V, depending on the 555 state
4	RESET	Keep this HIGH to prevent resetting timer interval

Number	Name	Purpose
6	THRES	If pin 6 (THRES) goes above 6 V, turns pin 3 (OUT) to 0 V
8	VCC	9 V, goes to the plus terminal of the battery

Capacitors

A capacitor (shown in Figure 2-17) stores electricity in an electric field. It has two conductive plates, separated by an insulator. Even the symbol for a capacitor shows this: two lines for the metal plates are separated by a gap.

Figure 2-16. *Three capacitors*

Figure 2-17. *Symbol for capacitor*

Capacitors are used in *resonant circuits*, like in radios. They can also *smooth* spikes in DC power sources (e.g., in the circuitry of an Arduino board). Yet another use is timing events, which is the usage you just worked with in the last project.

The capacitor and resistor you hook up to the 555 timer configure its mode. The values also further affect the specific expression of the configured mode and allow us to make determinations about the overall circuit.

How to Approach an Unfamiliar IC

When you see an IC, how are you supposed to know what it does? Here's something that usually works: read the tiny text on the chip and type it into a search engine. For example, some 555 timers are labeled CA555E and some are labeled NE555. The text on the chip doesn't need to make sense. In fact, the weirder the combination of letters, the more specific hits you'll get.

The datasheet tells you how a chip is used, so you can add that as a search term: "CA555E datasheet" or "NE555 datasheet." You don't need to understand the internals of an IC to use it, and datasheets of more complex chips only explain the connections of the chip. You could also try your luck for step-by-step instructions and search for "CA555E tutorial" or "NE555 tutorial."

These 555 projects wouldn't have been easy to accomplish without a 555 timer. A similar complexity simplification occurs when we move from the 555 timer to a microcontroller, and then again to a system-on-a-chip.

Firefly

Time to build the final circuit for the project. If you skipped here to just build the firefly—good, you can read the theory when it works. If you did every intermediate step, great—you can just combine them with the understanding you have built.

Build the circuit (see Figure 2-18). It simply combines the circuits you have already built: "Fading an LED" on page 24, "555 Fading Blink" on page 25, and "Light to Darkness" on page 21.

Parts

You need the following parts for this project:

- Two BC547 transistors
- 470 Ω resistor (four-band resistor: yellow-violet-brown; five-band resistor: yellow-violet-black-black; the last band will vary depending on the resistor's tolerance)
- 10 kΩ Resistor (four-band resistor: brown-black-orange; five-band resistor: brown-black-black-red; the last band will vary depending on the resistor's tolerance)
- 5 mm red LED
- 9 V battery
- 9 V battery clip
- Breadboard

- CA555E, NE555, or compatible 555 timer integrated circuit
- Potentiometer
- 100 uF capacitor
- Jumper wires
- Light-dependent resistor (LDR)

All of these parts, except the 9 V battery, 470 Ω resistor, and BC547, are available in the Maker Shed Mintronics: Survival Pack (*http://bit.ly/mintron-sp*), part number MSTIN2.

Build It

Build the circuit as shown in Figure 2-18.

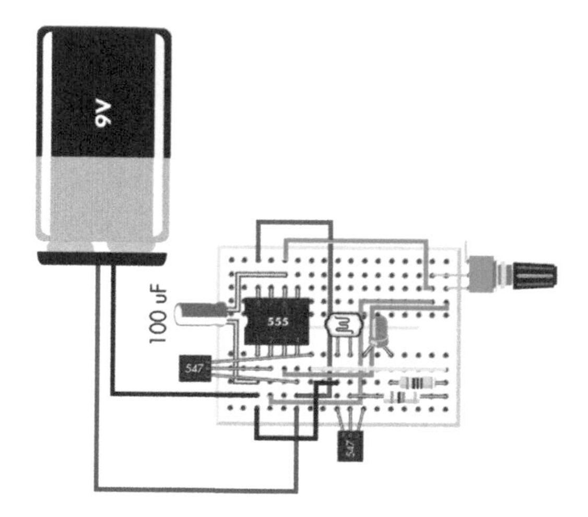

Figure 2-18. *Completed firefly*

Once the circuit is ready, try it out. Cover the LDR with your finger, and the LED gradually lights up, then dims, then lights up again. Point a flashlight at the LDR, and the LED turns off.

Troubleshooting

If this project doesn't work, try checking the following things:

- Double-check the connections.

- Test your LED and battery the way you learned at the beginning of this chapter.
- If you can't find out what's wrong, try the pulsating and photoresistor circuits separately again. This way it's easier to isolate the problem.

What's Next?

Congratulations, you have completed the chapter on component-by-component electronics! The next chapter is about Arduino, so things are just going to get easier. But before you move on, continue experimenting with the analog, component-by-component approach:

- Can you make your firefly light up in the day instead of night?
- Try replacing the potentiometer with a suitable resistor.
- Try replacing the LDR with any analog resistance sensor, so your firefly can react to being squeezed, the temperature, or the amount of alcohol on your breath.
- Try replacing the output with another output, such as a DC piezo speaker.

Enjoy your firefly!

3/Sensors and Arduino

In this chapter, you will learn how to process sensor data using an Arduino, the wildly popular microcontroller development board. There are many advantages to using Arduino rather than individual components as you did in Chapter 2. One of the biggest is that you'll save time not having to wire up complex circuits to process sensor measurements (see Figure 3-1).

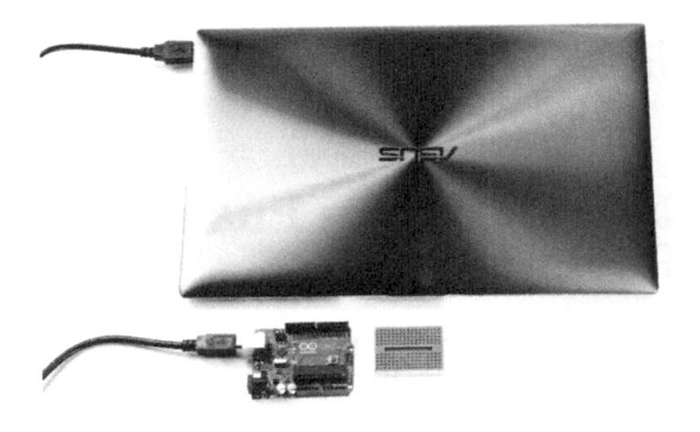

Figure 3-1. *Our Arduino Uno, breadboard, and laptop*

An Arduino runs *sketches*, which are programs that you write and upload to the Arduino from your computer. By writing software programs that make the decisions necessary to process sensor data, you really expand how much you can do, and, amazingly, you also cut down the time you'd have to spend working with hardware components. Sure, you have to write a few lines of code, but it's faster to write a conditional loop in code than to design it in hardware—at least for most of us.

It is also much easier to modify code than it is to modify a circuit layout. For example, suppose you want to add a delay how often a sensor takes measurements. You could accomplish this by chaining a few capacitors in series, but you'll need to do some careful planning and calculation to get the timing right. It's much easier to use a simple circuit and use the Arduino `delay()` function to control the timing. It is more efficient to type a few lines of code and upload a compiled binary program than it is to hunt for a specific component and wire it into a circuit. For example:

```
delay(1000);  // delays code execution for 1000 milliseconds or 1 second
```

Don't worry if you've never programmed in your life; you'll learn the basics in no time, and the more you tweak your code, the more you will learn. Sketch writing will open a whole new world of opportunities in your electronics projects. Time to get building!

If this is your first time using Arduino, read Appendix B to learn how to install drivers and the Arduino integrated development environment (IDE). It's also important to try the most simple program ("Hello, world") with your new hardware.

As long as the sensors and other components are used within their electrical specifications, using an Arduino or any other programmable microcontroller board will simplify your circuit and facilitate greater flexibility. However, if you use them outside of their specifications, such as applying more voltage to an LED than you should, you risk destroying the components, or even your Arduino.

Project 6: Momentary Push-Button and Pull-Up Resistors

Buttons are not the most exotic components but surely one of the most used and common sensors. Unlike some techniques you'll see for wiring a button to an Arduino, this project does not require an external resistor. That's because you'll be using the pull-up resistor that's internal to the Arduino's general-purpose input/output (GPIO) circuitry that can actually be turned on by a single line of code. Hardware doesn't get much more programmable than this.

Parts

You need the following parts for this project:

- Momentary push button
- Arduino Uno
- Jumper wires

- Breadboard

Build It

Here are the steps for building this project:

1. Orient your breadboard so that it is wider than it is tall, as shown in Figure 3-2.
2. Insert the momentary push button anywhere on the breadboard.
3. In a breadboard column with a button lead in it, plug in a jumper wire and connect the wire's other end to GND on the Arduino.
4. Plug another jumper wire into the same row as the jumper you just connected, but in a column corresponding to the button's other pair of leads. Next, attach the wire's other end to pin 2 on the Arduino.

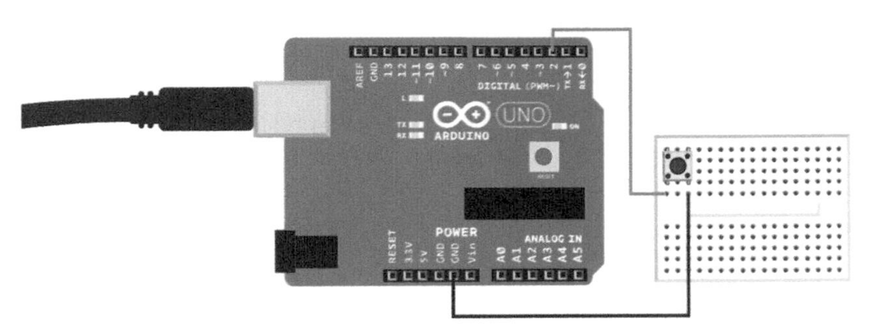

Figure 3-2. *Momentary push button connected to an Arduino Uno*

Run the Code

Once you're done hooking everything up, run the code shown in Example 3-1. When the button is pushed, the code lights up the built-in LED attached to pin 13.

Example 3-1. Arduino sketch for detecting a button press

```
// button.ino - light an LED by pressing button
// (c) BotBook.com - Karvinen, Karvinen, Valtokari

int buttonPin=2; // ❶
int ledPin=13;  // ❷
int buttonStatus=LOW;   // ❸

void setup()    // ❹
{
```

```
  pinMode(ledPin, OUTPUT);       // ❺
  pinMode(buttonPin, INPUT); // ❻
  digitalWrite(buttonPin, HIGH); // internal pull-up // ❼
}

void loop() // ❽
{
  buttonStatus=digitalRead(buttonPin); // ❾
  if (LOW==buttonStatus) { // ❿
    digitalWrite(ledPin, HIGH); // ⓫
  } else {      // ⓬
    digitalWrite(ledPin, LOW);  // ⓭
  }
} // ⓮
```

❶ Set up some *variables*. From now on, when you write `buttonPin` in your sketch, Arduino will put the number 2 in its place. Variables make code easier to read and modify because you can change them in a single place. All variables in this sketch are integers (`int`): whole numbers such as 1, 2, 3, or -500.

❷ Create another variable for the LED pin. The built-in LED is on digital pin 13 (`D13`).

❸ Even though the value of `buttonStatus` is modified elsewhere in the program, it's a good idea to initialize it to some value rather than leaving it undefined.

❹ The `setup()` function runs once when Arduino boots up. It contains all the one-time initialization you need in your sketch.

❺ Setting a pin to `OUTPUT` allows you to control it with the `digitalWrite()` function later.

❻ We set `buttonPin` (digital pin 2) to INPUT `mode`. This way, we can later read its value with `digitalRead()`.

❼ Enable the internal pull-up resistor. Magic (hardware being controlled by software) happens here! You'll learn some pull-up resistor theory in "Pull-Up Resistors and Arduino" on page 37. Note that you can accomplish the same thing by replacing this line and the preceding line with `pinMode(buttonPin, INPUT_PULLUP);`, but this will only work with recent versions of the Arduino software; it is not supported by older versions.

❽ Arduino runs the `loop()` function automatically after `setup()` has finished. As the name implies, `loop()` is called again and again until you reset or power down the Arduino.

9. Here you read the status of `buttonPin` (which you defined earlier in the sketch as digital pin 2). Then you save this answer to variable `buttonStatus`. If the button is pressed (*closed*), pin 2 is connected to ground and digitalRead returns `LOW`. If the button is not pressed (*open*), pin 2 is not connected to GND, so the internal pull-up resistor pulls D2 up to +5 V, returning HIGH.

10. Two equals signs (==) indicate a comparison between two expressions. Note that this is in contrast to a single equals sign (=), which is used for assigning a value to a variable. Always use two equals signs when performing a comparison in an `if` statement. By putting `LOW` on the left side of the comparison, you guard against such an error: LOW is a constant, and you cannot assign a value to it. So if you use a single equals sign, Arduino would give you an error message when you try to compile and load the sketch to your Arduino.

11. Here the LED is illuminated.

12. The `else` clause handles the case in which the `if` comparison doesn't evaluate to true. The block of code ({}) after the `else` is executed when `LOW != buttonStatus` (which is equivalent to `buttonStatus != LOW`, and means `buttonStatus` is not equal to `LOW`).

13. Turn off the LED.

14. As soon as `loop()` finishes, Arduino runs it again, starting at the top of the `loop()` function.

You can download all source code from this book's website (*http://getstarted.botbook.com*).

Pull-Up Resistors and Arduino

The momentary push-button design works reliably because you engaged an internal *pull-up resistor* in the Arduino circuitry. You might wonder exactly what a pull-up resistor pulls: it pulls the voltage of the pin toward the 5 volts that power the Arduino.

Never trust a floating pin. If you read a pin that's not connected to anything, you get an unpredictable answer. For example, if you don't connect anything to digital pin D2 and then read it with `digitalRead()`, there are no guarantees as to what you'll get as a result. It could be `HIGH`, it could be `LOW`, it could change 10 times a second or never change. A floating pin gives a useless value (see Figure 3-3).

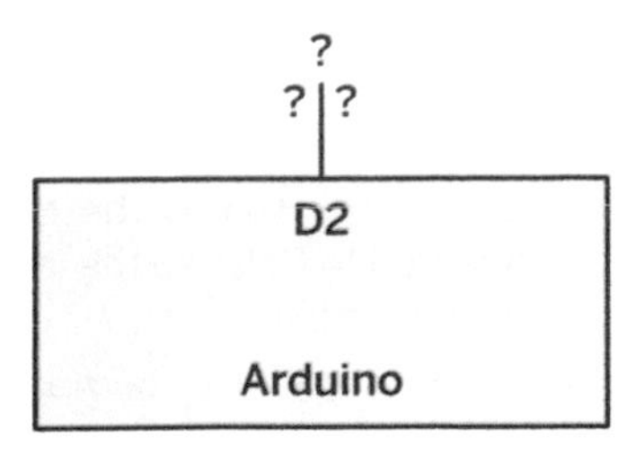

Figure 3-3. *There is no point in reading a floating pin; add a pull-up resistor*

The Arduino IDE comes with many useful example sketches. If you look closely at their button code (File → Examples → 02.Digital → Button) it actually requires an external resistor and won't work with our circuit. That's because that sketch uses a *pull-down* resistor, so a HIGH state indicates that the button is closed (pressed) and a LOW state indicates that the button is open.

If you connect the same D2 to GND (zero volts, ground, LOW), your `digital Read()` is guaranteed to return a result of LOW (see Figure 3-4). This handles the case where the button is pressed.

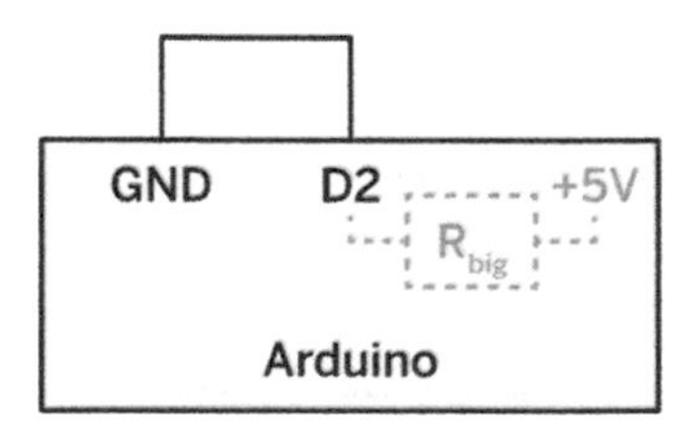

Figure 3-4. *LOW; pressing the button between D2 and GND is the same as connecting D2 to GND*

What if you *don't* press the button? We already know you have to connect a pin somewhere to read its value. But even if you connect a button to Arduino, the pin isn't really connected to anything until you press it.

Pull-up resistor to the rescue!

The pull-up resistor connects D2 to +5 V (HIGH), albeit weakly (because the resistor limits the current). If nothing else is connected to D2, the resistor will pull it up to HIGH. So when the button is open, D2 is not connected to ground, but instead to +5 V through the pull-up resistor, which means that D2 will be HIGH until you press the button (see Figure 3-5).

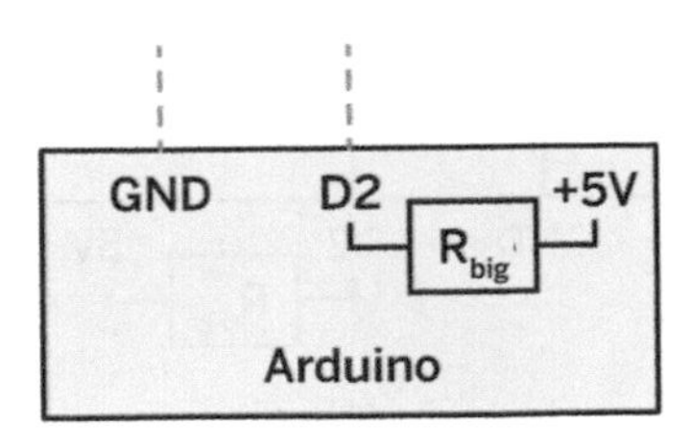

Figure 3-5. *HIGH; button up, D2 is only connected to +5 V through the pull-up resistor*

Now what happens when you press the button? It's clear that D2 and GND are now connected, but so is +5 V, isn't it? You may be wondering, "You are not telling me to connect GND and +5 V, are you? Isn't that a short circuit?"

Although GND and D2 are both connected to +5 V, that connection is through a big resistor, and because the path between GND and D2 has *less* resistance than the resistor that connects to +5 V, the current in the circuit flows through GND and D2. In other words, the pull-up resistor pulls D2 up to HIGH (+5 V) if nothing else is pulling it down.

Typical values for pull-up resistors are tens of thousands of ohms to a few million ohms (e.g., 20 kΩ to 2 MΩ).

This is how your connection works: press the button, D2 goes LOW. Release the button, and the pull-up resistor pulls it up to HIGH. How do you recognize a floating pin in your own connection? If any sensor connects to Arduino with just two leads, ask yourself whether the pin is connected in every possible state the sensor can be in. If not (as in the case with a button that's open), you probably need a pull-up resistor (see Figure 3-6) to hold it HIGH when it's not being pressed.

If you prefer to use an external resistor, it's more common to use it as a pull-down to hold the pin LOW. If you do that, remember to reverse the logic in your code (HIGH indicates a button press, LOW indicates that the button is open).

That's it for the diagrams and theory for now. As you can see, even the humble switch has a lot to teach us about how sensors need to be wired up. With that background out of the way, it's time to relax with some practical, hands-on stuff.

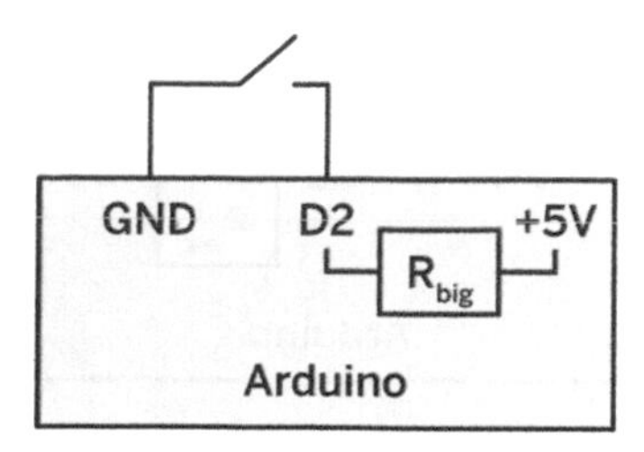

Figure 3-6. *Internal pull-up resistor; enable it with pinMode(pin, INPUT) followed by digitalWrite(pin, HIGH) or just pinMode(pin, INPUT_PULLUP)*

Project 7: Infrared Proximity to Detect Objects

Infrared sensors (such as the switch shown in Figure 3-7) consist of two parts: an infrared emitter and receiver. The emitter is actually an LED that emits light that's invisible to the human eye, as the wavelength is longer than light in the visible spectrum. An infrared LED is the same component that is used in a TV remote control.

The receiver part of the switch collects the IR light that is reflected back. Obstacles cause more light than usual to be reflected, which tells you that there is something in front of the sensor.

This sensor can become confused in strong sunlight, and some very dark materials can pass by unnoticed.

Figure 3-7. *Infrared sensor switch*

You can buy a switch like this from DFRobot (part #SEN0019), as well as from sellers on Amazon, eBay, AliExpress, and DealeXtreme. They are usually listed as "Adjustable Infrared Sensor Switch" or under the manufacturer part number E18-D80NK.

An infrared sensor switch is useful for various projects. It's especially handy when you need to know if something is close, but when you don't need to know the object's exact distance. You can set the detection range of the sensor by adjusting the trimmer potentiometer (trimpot) with a small screwdriver, as shown in Figure 3-8.

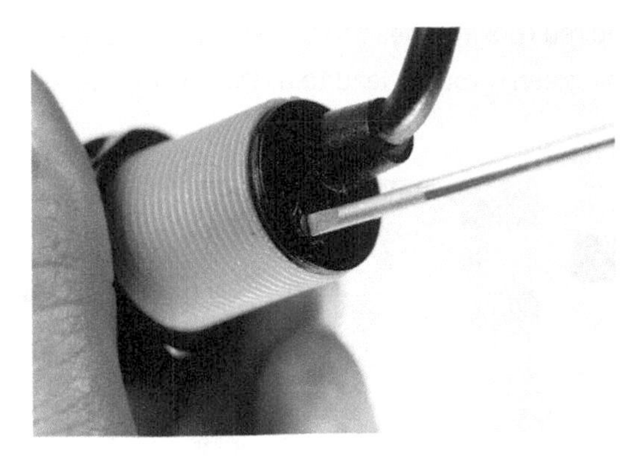

Figure 3-8. *Adjusting the infrared sensor*

Parts

You need the following parts for this project:

- Infrared sensor switch
- Arduino Uno
- Jumper wires

If your infrared switch has wires that terminate in female headers, you'll need to use jumper wires to connect each wire to the Arduino. If it terminates in male headers, you may need to build this circuit using a breadboard and jumper wires.

As long as you have a sensor that breaks out wires for ground, 5 V, and a signal, then this layout will work with your sensor. We used a sensor with separated wires, so we inserted each lead directly into the Arduino headers.

Build It

Make sure the Arduino is powered down. Connect the circuit as shown in Figure 3-9, then run the sketch shown in Example 3-2. When you put something in front of the sensor, the built-in LED attached to pin 13 is lit.

Follow these instructions for connecting the circuit:

1. Connect the yellow (signal) lead to digital pin 8 on the Arduino.
2. Connect the red (positive) lead to the +5 V pin on the Arduino.
3. Connect the green (ground) lead to a GND pin on the Arduino.

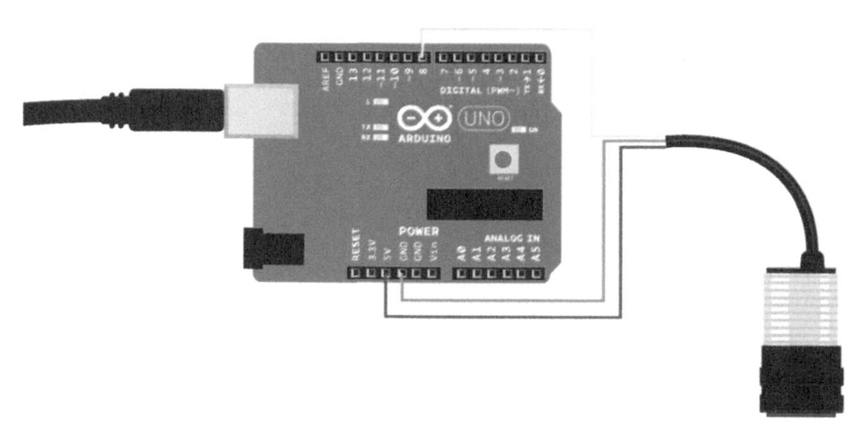

Figure 3-9. *Connecting the infrared sensor switch*

Example 3-2. Infrared distance switch Arduino sketch

```
// infrared_proximity.ino - light LED when object near, using Dagu IR switch
// (c) BotBook.com - Karvinen, Karvinen, Valtokari

int irPin=8; // ❶
int ledPin=13;
int objectDetected=LOW; // ❷

void setup() {
  pinMode(ledPin, OUTPUT);
  pinMode(irPin, INPUT);
  digitalWrite(irPin, HIGH); // internal pull-up
}

void loop() {
  objectDetected=digitalRead(irPin);
  if (LOW==objectDetected) {
    digitalWrite(ledPin, HIGH);
  } else {
```

```
    digitalWrite(ledPin, LOW);
  }
}
```

❶ This is similar to the code from Example 3-1, but the name `buttonPin` is changed to `irPin`.

❷ Similarly, the name `buttonStatus` is changed to `objectDetected`.

You know this code already! As you look at it, you might recognize that it's the same code you used back in Example 3-1. We changed two variable names, but that is largely cosmetic. Experiment with it yourself: can you use the infrared proximity sensor with the code from Example 3-1? Can you use a button with the code from Example 3-2?

Many sensors have identical interfaces. Now that you know how to work with two switches, which we refer to as *digital resistance sensors*, you can apply this skill to similar sensors in your own projects.

Analog Resistance Sensors and Voltage Dividers

Most sensors you'll use are analog resistance sensors. They produce their results with a gradual change in resistance. For example, the more force you exert on a force-sensing resistor like a FlexiForce, the lower its resistance.

Because you can't measure resistance directly with Arduino (or Raspberry Pi), you'll learn how to use these types of sensors with an external resistor to form a *voltage divider*. A voltage divider looks a lot like a pull-up or pull-down resistor configuration, but instead uses two resistors in series between +5 V and GND:

- The external resistor
- The sensor, whose resistance varies depending on its state

By using an analog pin to measure the voltage between these two resistors, you'll receive a voltage that varies in proportion to the state of the sensor.

Project 8: Rotation (Pot)

The potentiometer (shown in Figure 3-10) is an adjustable resistor with a knob. When you turn the knob, its resistance changes. Potentiometers are very common in everyday items such as radios and clothes irons. In RC airplanes, servomotors can determine their angular position with the help of a potentiometer.

Figure 3-10. *Potentiometers can measure rotation*

To some extent, analog resistance sensors can behave like digital ones, provided they meet two criteria:

- At their lowest level of resistance, they must let enough current through to register as HIGH.
- At their highest level of resistance, they must resist enough current to register as LOW.

Try it yourself! Return to "Project 6: Momentary Push-Button and Pull-Up Resistors" on page 34, hook everything up, and then run the code. Then, replace the button with the potentiometer: one side of the potentiometer to +5 V, the other to GND, and the middle pin to digital pin 2. Now, turn the knob to max or min. If the resistance of the potentiometer is sufficient, you can turn the LED on and off.

Because the code is designed to read a button, you can only turn the LED on or off. To make use of the analog values produced by the potentiometer, you must use the `analogRead()` function. When used with analog sensors, this function gives you a range of values (from 0 to 1023), rather than just HIGH or LOW.

This project controls an LED's blinking speed with a potentiometer. Turn it all the way in one direction, and the LED blinks slowly. Turn it all the way in the other direction, and it blinks faster.

Figure 3-11 shows the wiring diagram for this project.

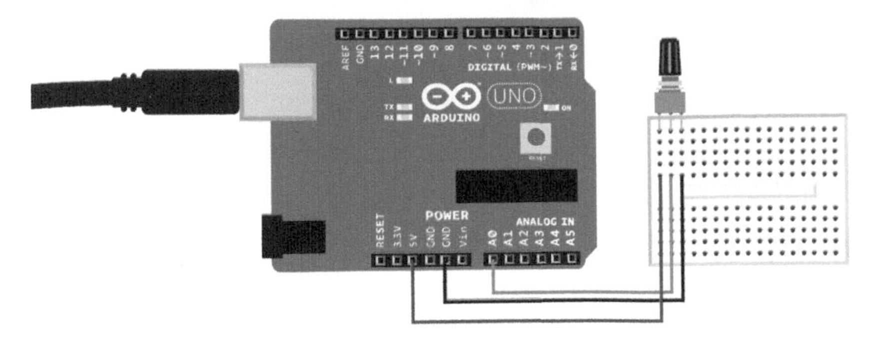

Figure 3-11. *Potentiometer wired to GND and A0 of an Arduino Uno*

The potentiometer acts as a voltage divider. One side is connected to +5 V, and the other to GND. The center pin is attached to analog input pin A0. As you turn the knob, the resistance on either side of the pin varies. For example, for a 10 kΩ potentiometer (all numbers approximate because inexpensive potentiometers are not 100% precise):

When the knob is turned all the way to one side[1]

- The resistance between the positive and center pin is zero.
- The resistance between the center and negative pin is 10 kΩ.
- Pin A0 receives a voltage of 5 V.
- `analogRead(potPin)` returns 1023.

When the knob is all the way to the other side

- The resistance between the positive and center pin is 10 kΩ.
- The resistance between the center and negative pin is zero.
- Pin A0 receives a voltage of 0 V.
- `analogRead(potPin)` returns 0.

When the knob is centered

- The resistance between the positive and center pin is 5 kΩ.
- The resistance between the center and negative pin is 5 kΩ.
- Pin A0 receives a voltage of 2.5 V.
- `analogRead(potPin)` returns approximately 512.

Parts

You need the following parts for this project:

1. Which side depends on how you insert the pot into the breadboard; try it both ways!

- Potentiometer (around 10 kΩ recommended)
- Arduino Uno
- Jumper wires
- Breadboard

Build It

Wire the components up as shown in Figure 3-11.

Run the Code

Next, run the code listed in Example 3-3.

Example 3-3. Arduino sketch for reading a potentiometer

```
// pot.ino - control LED blinking speed with potentiometer
// (c) BotBook.com - Karvinen, Karvinen, Valtokari

int potPin=A0; // ❶
int ledPin=13;  // ❷
int x=0; // 0..1023 // ❸

void setup() {  // ❹
  pinMode(ledPin, OUTPUT);       // ❺
}

void loop() {   // ❻
  x=analogRead(potPin); // ❼
  digitalWrite(ledPin, HIGH);   // ❽
  delay(x/10); // ❾
  digitalWrite(ledPin, LOW);    // ❿
  delay(x/10);  // ⓫
} // ⓬
```

❶ In earlier examples, we used Arduino's digital pins, and we referred to them just by number (such as `13` for digital pin 13). The analog input pins are also numbered, but we put the letter `A` in front of the name when referring to them (`A0`, `A1`, etc.). The analog pins are on the opposite side of Arduino's digital pins.

❷ Digital pin 13 is connected to the Arduino's built-in LED.

❸ `x` holds raw values that `analogRead()` returns. Note that we use the comment to show the expected range of values. Even though you can find this information in the Arduino documentation, it's handy to have here.

❹ `setup()` performs one-time initialization.

❺ Set the pin D13 to OUTPUT, so we can control it later with `digitalWrite()`. Note that although you'll be reading A0 later, it's not necessary to explicitly configure it as an INPUT, because `analogRead()` behaves differently than `digitalRead()`.

❻ After `setup()` is finished, `loop()` is called automatically.

❼ Read the voltage from `potPin` (A0) with `analogRead()`. You'll receive a value between 0 (0 V == LOW) and 1023 (+5 V == HIGH). The code saves this value to variable `x`.

❽ Turn on the onboard LED.

❾ We'll wait for x/10 milliseconds. That's 0 when the knob is turned to the max (A0 reads 0 V). When the knob is turned the other way, the delay is 0.1 seconds (1023/10 ms).

❿ Turn off the onboard LED.

⓫ Wait again, so that the LED remains off briefly.

⓬ After `loop()` finishes, it's called automatically again.

Project 9: Photoresistor to Measure Light

How bright is it? A photoresistor lets more electricity through when it's in bright light (see Figure 3-12). Our students have used photoresistors to make robots love (or hate) light, activate lights in the dark, or create burglar alarms.

Figure 3-12. *A photoresistor is a better conductor in bright light*

Parts

You need the following parts for this project:

- Photoresistor (10 K recommended)

- 10 kΩ resistor (four-band resistor: brown-black-orange; five-band resistor: brown-black-black-red. The fourth or fifth band will vary depending on the resistor's tolerance)
- Arduino Uno
- Jumper wires
- Breadboard

Build It

Figure 3-13 shows the circuit design for this project. Build it as shown, and run the sketch listed in Example 3-4. Note the addition of the 10K resistor. This is to create a voltage divider. Back in "Project 8: Rotation (Pot)" on page 43, you didn't need a separate resistor because the potentiometer can be used as a voltage divider.

In order to produce a voltage that varies in proportion to the photoresistor's change in resistance, you need to create a voltage divider with the two resistors (the 10K resistor, and the photoresistor) and measure the voltage that's produced between them to determine how dark or light it is.

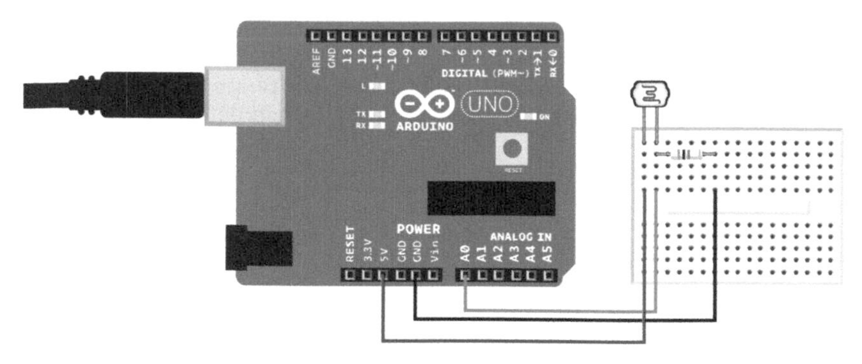

Figure 3-13. *Photoresistor circuit layout*

Run the Code

Example 3-4. Arduino photoresistor sketch

```
// photoresistor.ino - blink faster in dark, slower in the light
// (c) BotBook.com - Karvinen, Karvinen, Valtokari

int photoPin=A0;
int ledPin=13;
int x=-1; // 0..1023

void setup() {
  pinMode(ledPin, OUTPUT);
}

void loop() {
  x=analogRead(photoPin);
  digitalWrite(ledPin, HIGH);
  delay(x/10); //
  digitalWrite(ledPin, LOW);
  delay(x/10);
}
```

Hey, wait a minute; this looks a lot like Example 3-3! It's exactly the same code. We did replace "potPin" with "photoPin" and changed the comment on the first line. Run the program and try it out. What happens when you make a shadow or hide the photoresistor completely?

Most common sensors are analog resistance sensors, so they use a variation of this same connection. You can apply this same approach to similar sensors you meet in the future.

Project 10: FlexiForce to Measure Pressure

A force-sensitive resistor such as the FlexiForce (see Figure 3-14) is a flat sensor that measures how much pressure you exert on it. Untouched, without any pressure, it has a very high resistance. When you press the round area at the end of the sensor, the resistance drops and it becomes more conductive.

Figure 3-14. *A force-sensitive resistor measures pressure exerted on it*

Parts

You need the following parts for this project:

- Force-sensitive resistor (FSR)
- 1 MΩ resistor (four-band resistor: brown-black-green; five-band resistor: brown-black-black-yellow. The fourth or fifth band will vary depending on the resistor's tolerance)
- Arduino Uno
- Jumper wires
- Breadboard

Build It

You'll use a voltage divider here just as you did in "Project 9: Photoresistor to Measure Light" on page 47. As with that project, you need to use a resistor that's in the same range as the FlexiForce. We found that a 1 MΩ resistor produced suitable results.

In our voltage divider, analog pin A0 measures the variation in voltage ("Analog Resistance Sensors and Voltage Dividers" on page 43) as you change the amount of force exerted on the sensor, which gives you a value between 0 (voltage is near GND) to 1023 (voltage is near +5 V). Build the circuit as shown in Figure 3-15, and then run the sketch from Example 3-5.

Watching the Action Over the Serial Port

You don't need to guess what's going on inside your Arduino's brain. Arduino can talk to your computer over its built-in serial port and send text back to it (you can even send messages from the computer to Arduino), as shown in Figure 3-16. The Arduino uses its USB port to transfer serial messages.

Serial ports are a very common way for embedded devices to talk. With our book on controlling robots with your brainwaves (*Make a Mind-Controlled Arduino Robot*), we used a serial-over-USB cable to hack a commercial, off-the-shelf EEG headband. This allowed us to control the Arduino-based robot with our brainwaves! Another great example is our mobile phone controlled football robot from our book, *Make: Arduino Bots and Gadgets*. In that project, we used serial over Bluetooth, allowing the Arduino to communicate with an Android phone. But serial is not just used for communicating directions to devices. When building an early prototype of our satellite, we used serial-over-USB to debug our project. In fact, that's what we'll do in this project, too.

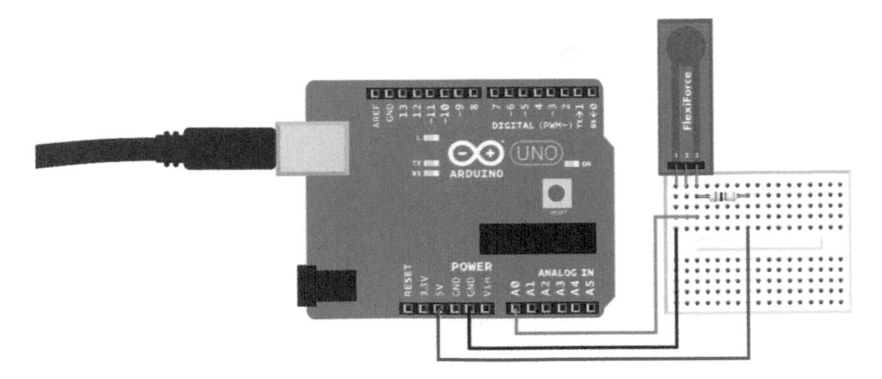

Figure 3-15. *FlexiForce circuit layout*

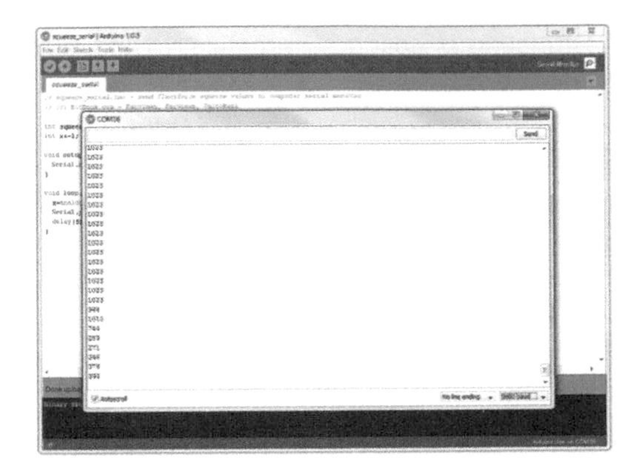

Figure 3-16. *Serial Monitor*

Run the Code

Example 3-5. Arduino sketch for reading the FlexiForce sensor

```
// squeeze_serial.ino - flexiforce squeeze level to serial
// (c) BotBook.com - Karvinen, Karvinen, Valtokari

int squeezePin=A0;       // ❶
int x=-1; // 0..1023  // ❷

void setup() {  // ❸
  Serial.begin(9600); // bit/s // ❹
}

void loop() {   // ❺
  x=analogRead(squeezePin);  // ❻
```

```
  Serial.println(x); // ❼
  delay(500); // ms     // ❽
}
```

❶ This is the pin that reads the sensor (analog pin 0).

❷ Variable `x` is global; it's declared outside any functions, so it would be available to any other functions you might add to your sketch. We initialize it to a value (-1) that you'd never get from `analogRead()` to make debugging easier: if you see this value in the results, then you would know there's a problem.

❸ The `setup()` function is called automatically; it handles one-time initialization.

❹ The serial port speed (we chose 9600 because that's what the Arduino Serial Monitor uses by default; if you've changed the setting on the Serial Monitor, you will need to change it back to 9600).

❺ The `loop()` function is called automatically (and repeatedly) after `setup()` is finished.

❻ `analogRead()` returns a value between 0 (LOW, 0 V) and 1023 (HIGH, +5 V). This value is stored to variable `x`.

❼ This prints out the value of the variable to the serial port so that it appears in the Serial Monitor. `Serial.println()` prints each value on its own line, so when you open the Serial Monitor, you'll see each value scrolling past in the window.

❽ Wait for a while: this gives you time to observe each value in the Serial Monitor, and prevents your code from consuming 100% of the Arduino's CPU time.

Experiment yourself: can you connect a potentiometer and read its current value through serial? If you need to challenge yourself, you can try converting rotation to angle in degrees or a percentage of maximum rotation.

Are you feeling really adventurous? Once you get your data to your computer, you can read it in any programming language. In our interactive painting from our book *Make: Arduino Bots and Gadgets* (*http://bit.ly/make-abg*), we use Python's serial library to read values from the Arduino and control images on a computer.

Project 11: Measuring Temperature (LM35)

Is it hot in here? The LM35 is an inexpensive thermometer that's also easy to use (see Figure 3-17). You just need to retrieve the voltage with `analog Read()`, and then you can calculate the temperature in degrees Celsius as `voltage * 100C/V`.

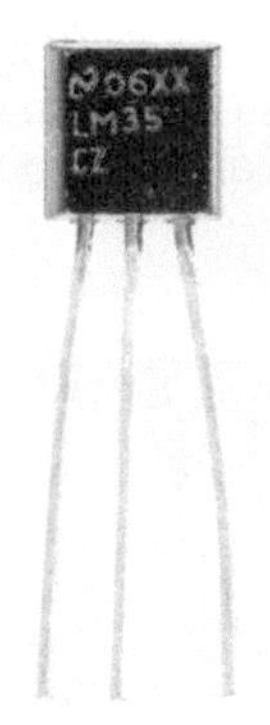

Figure 3-17. *LM35 temperature sensor*

Our students have used the LM35 to measure temperature in their summer cottages and greenhouses, for example.

Parts

You need the following parts for this project:

- LM35 temperature sensor
- Arduino Uno
- Jumper wires
- Breadboard

The LM35 differs from previous analog resistance sensors you have used. It has three leads: 5 V, ground, and signal. In this way, it's easier to wire than others because you don't need a pull-up resistor.

Build It

Figure 3-18 shows the circuit diagram. Build it as shown, and then run the code listed in Example 3-6. Open the Serial Monitor to see the current temperature. You'll need to follow these steps:

1. Connect the LM35 sensor to Arduino using a breadboard. If you're using the most common version of the LM35 (the TO-92 package, which is a common semiconductor package for transistors), hold the leads down

and orient the flat side toward you. The left lead is +5 V, middle is VOUT, and the right lead is GND.

2. Connect +5 V to Arduino's +5 V and GND to Arduino's GND. But what is VOUT? Voltage OUT indicates the temperature as an output voltage, so you need to connect it to an analog input pin on Arduino (Example 3-6 uses A0).

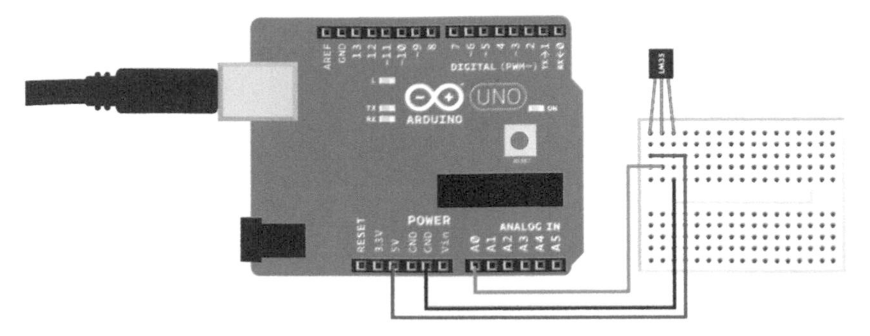

Figure 3-18. *Temperature sensor connected using breadboard*

Once you've measured the room temperature, you might want to test the LM35 some more. If you put it in your fridge, prepare to wait. Air is a very good insulator for heat and doesn't conduct well. When we want to quickly test cooler temperatures, we use ice cubes (see Figure 3-19). Wet ice cubes quickly conduct heat away from the LM35, so we get instant readings.

Figure 3-19. *Testing the LM35 with an ice cube*

Run the Code

Example 3-6. The LM35 Arduino circuit

```
// temperature_lm35.ino - LM35 temperature in Celsius to serial
// (c) BotBook.com - Karvinen, Karvinen, Valtokari

int lmPin = A0; // ❶

void setup()    // ❷
{
  Serial.begin(9600);   // ❸
}

float tempC()   // ❹
{
  float raw = analogRead(lmPin);  // ❺
  float percent = raw/1023.0; // ❻
  float volts = percent*5.0; // ❼
  return 100.0*volts;  // ❽
}

void loop()     // ❾
{
  Serial.println(tempC());  // ❿
  delay(200); // ms     // ⓫
}
```

❶ We use analog pin 0 (A0) to read the output of the sensor.

❷ When the Arduino boots up, `setup()` gets called automatically to perform one-time initialization.

❸ Opens the serial port. You can monitor the Arduino's serial port output by going to Arduino → Tools → Serial Monitor. Be sure to select the same speed in both this code (it's 9600 here) and in the Serial Monitor.

❹ Define a new function called `tempC()`. You'll call this function later in the code when you need to read the temperature.

❺ The function `analogRead()` returns an integer value from 0 to 1023. A value of 0 means 0 V (LOW), and a value of 1023 means +5 V (HIGH).

❻ We convert the raw reading (0-1023) to a percentage by dividing it by the maximum value. The percentage will be 0.0 for 0% and 1.0 for 100%. To get the floating-point (decimal) answer you need, you must use the floating-point divisor 1023.0 instead of integer 1023. In C and C++, 1/2==0 (integer division), but 1/2.0==0.5 (floating-point division).

❼ The `percent` variable is calculated out of a maximum of 5.0 V (HIGH). So multiplying by 5.0 gives the voltage in volts.

❽ From the LM35 datasheet (search the Web for "lm35 datasheet"), you can see that the scaling factor is 10 mV/C, with 0 V meaning 0 °C. Playing with the math, 10 mV/C = 0.010 V/C. The inverse is 1/0.010 C/V = 100 C/V. So you can get the temperature in °C by multiplying the voltage by 100.

❾ After `setup()` finishes, `loop()` is called automatically, over and over until you power down the Arduino.

❿ Call the `tempC()` function, which returns a number (e.g., 20). Print this number and a line break to the Serial Monitor. Writing `tempC()` as a function makes it easy to use it in our other projects. Now that it works, you don't have to think about the implementation details. When you want to know the temperature, just connect the LM35 and call `tempC()`.

⓫ Pause for a while. This gives you time to read the text, keeps the serial buffer from overflowing, and prevents the program from hogging 100% of the Arduino's CPU.

Project 12: Ultrasonic Distance Measuring (HC-SR04)

The ultrasonic distance sensor (shown in Figure 3-20) is one of our favorite sensors. It sends an ultrasonic sound and then listens for how quickly it returns, like a tiny sonar. As the name implies, the sound frequency is too high for human ears to hear. The great thing compared to the infrared sensor switch you saw earlier ("Project 7: Infrared Proximity to Detect Objects" on page 40) is that this sensor actually gives you the distance to the obstacle (you can find infrared distance sensors, such as the Sharp GP2Y0A series, that report distance as well).

Nothing is without trade-offs, so while you can get a distance reading, the ultrasonic distance sensor can miss certain objects easily. Items that are especially soft and inclined planes tend to go unnoticed. This is one reason that stealth aircraft are shaped the way they are.

There are many ultrasonic distance sensors. The Parallax PING used to be our favorite, but its cost of about $30 limited how many we could buy. Now there are ultrasonic distance sensors available for a couple of dollars each. The HC-SR04 is one of these.

Figure 3-20. *Ping ultrasonic distance sensor*

Parts

You need the following parts for this project:

- HC-SR04 ultrasonic distance sensor
- Arduino Uno
- Breadboard
- Jumper wires

Build It

Wire up the sensor as shown in Figure 3-21 and run the code in Example 3-7.

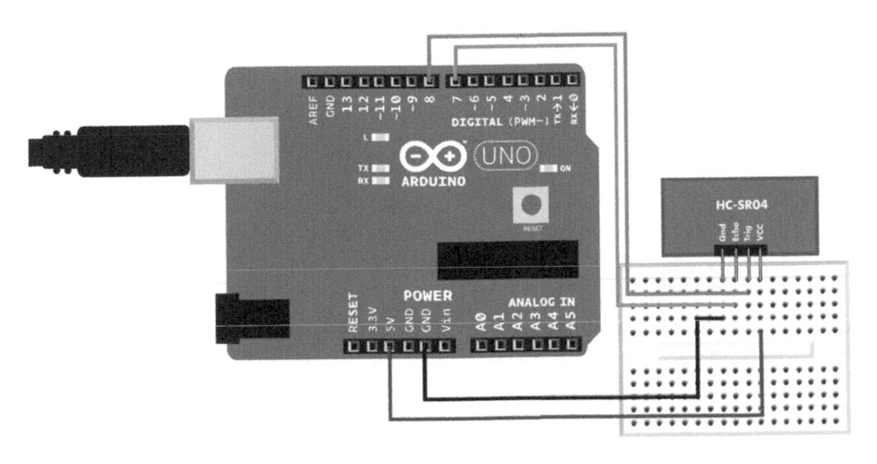

Figure 3-21. *HC-SR04 connected with breadboard*

Run the Code

Example 3-7. The HC-SR04 Arduino sketch

```
// hc-sr04.ino - distance using ultrasonic ping sensor
// (c) BotBook.com - Karvinen, Karvinen, Valtokari


int trigPin = 8;          // (1)
int echoPin = 7;          // (2)
float v=331.5+0.6*20; // m/s    // (3)

void setup()    // (4)
{
  Serial.begin(9600);   // (5)
  pinMode(trigPin, OUTPUT);      // (6)
  pinMode(echoPin, INPUT);       // (7)
}

float distanceCm(){      // (8)
  // send sound pulse    // (9)
  digitalWrite(trigPin, LOW);    // (10)
  delayMicroseconds(3); // (11)
  digitalWrite(trigPin, HIGH);   // (12)
  delayMicroseconds(5); // (13)
  digitalWrite(trigPin, LOW);    // (14)

  // listen for echo    // (15)
  float tUs = pulseIn(echoPin, HIGH); // microseconds    // (16)
  float t = tUs / 1000.0 / 1000.0 / 2.0; // s   // (17)
  float d = t*v; // m    // (18)
  return d*100; // cm    // (19)
}

void loop()      // (20)
{
  int d=distanceCm(); // (21)
  Serial.println(d, DEC);        // (22)
  delay(200); // ms      // (23)
}
```

(1) You use the trigger pin to tell the sensor to transmit the sound pulse. It is marked with "Trig" on the HC-SR04 board. It's connected to Arduino digital pin D8, and when you take the pin HIGH, the pulse will be triggered.

(2) The echo pin goes high when the pulse returns. On HC-SR04, it's marked as "Echo." On the Arduino, it's connected to digital pin D7.

❸ The speed of sound is about 340 meters per second. It's calculated here for 20 °C air temperature (we add `20 * .6` to `331.5`). We always spell out calculations in our code for clarity. It doesn't add any extra work to Arduino, because the compiler turns the calculation into a constant before uploading your code to the Arduino.

❹ The `setup()` function is automatically run once when Arduino boots.

❺ Open the serial connection so that the result can be displayed to the Serial Monitor (Tools → Serial Monitor). Remember to set the same speed (bit/second, or "baud") in both your code and the Arduino Serial Monitor.

❻ Set the Trig pin to OUTPUT so `digitalWrite()` can later turn it on and off (HIGH or LOW).

❼ Set the Echo pin to INPUT so that `digitalRead()` can later read it to find out whether it's HIGH or LOW.

❽ Define a new function. This function `distanceCm()` returns a float, a decimal number. It takes no parameters, so there is nothing inside the parentheses after the name.

❾ There are two separate parts to reading the distance. First, we ask the HC-SR04 to send the pulse, and later, we check to see how long it took for the echo to come back.

❿ To create a clean pulse for sending the sound, the pin must be off (LOW) first. Music starts from silence.

⓫ Wait a very short time for the pin to settle. When working with different multiples of units, you should convert them to base units such as seconds to get an idea of the scale. Three microseconds is three millionths of a second, or 0.000003 seconds.

⓬ Turn the trigger pin on (HIGH, +5 V). The HC-SR04 is now transmitting the ultrasound.

⓭ Wait for a very short time, 5 ms (microseconds). Five microseconds == five millionths of a second == 0.000005 seconds. The trigger pin stays high during this time.

⓮ Turn the trigger pin off, ending the short sound pulse.

⓯ Use a comment to mark that we're in a different part of the function.

⓰ Now you need to find out how long it takes for the echo pin to go LOW. The time returned by the `pulseIn()` function is reported in microseconds. It's usually best to use plain ASCII characters in source code, so we indicate micro (millionth) with a plain *u* or *U* instead of the Greek letter μ.

⑰ Convert microseconds (millionths) to seconds by dividing. It's often practical to use SI base units, such as seconds, meters, or kilograms. To get decimal numbers and avoid premature rounding, we use a floating point in the divider (1000.0 instead of integer 1000). The time *t* is a round trip (there and back, two way, ping and pong), so we divide it by two to get one-way time.

⑱ To get distance from time, we multiply by speed. For example, if it took us 2 hours to get to Helsinki and back at 100 km/h, we'd calculate that distance with `2h/2*100km/h`, which is 100 km.

⑲ To get centimeters, we multiply meters by 100. This is a convenient unit for distances typically measured with HC-SR04.

⑳ After `setup()` finishes, `loop()` is called automatically (over and over). The `loop()` function is what your Arduino is doing most of the time.

㉑ This is all you need in your main program to use HC-SR04. Because distanceCm() is written as its own function, you can easily add it to your own projects. Even though it required some thought to write, we can forget the implementation details when we use it. To get distance, just attach the HC-SR04 sensor, include the function in your code, and call `distanceCm()`. Here, we cast (convert) the result to an integer, ignoring the decimal part. So `d` will now contain the distance in full centimeters (e.g., 23).

㉒ Print the distance to the Serial Monitor (Tools → Serial Monitor). The command `println()` prints the value of `d` as a decimal number, then a line break.

㉓ When running in an infinite loop, always wait for a while to avoid taking 100% of the Arduino's CPU time. This also gives you time to read the results in the Serial Monitor so it's not scrolling by too fast.

Experiment yourself! Can you evade the ultrasound? What surface reflects sound so that distance measuring works? What surfaces absorb or misdirect the sound, so that you become invisible to this sonar?

Looking for more of a challenge? Modify the `distanceCm()` function to take the digital pin as parameter (e.g., `distanceCm(3)` for HC-SR04 on digital pin 3). You can even connect multiple HC-SR04 sensors to the same Arduino.

Many projects use PING, HC-SR04, and similar ultrasonic distance sensors. We have used them in an interactive painting and a walking robot insect. Our students have used them in robots, a gesture mouse, burglar alarms, and interactive fluffy toys.

Conclusion

You have learned to write Arduino programs and measure the world. Your Arduino can sense pressure, temperature, light, and many other things. If you want to keep working with the Arduino to build robots and design your own devices, see our other book, *Make: Arduino Bots and Gadgets*.

Because the Arduino is simple and robust, it's a tool we always keep on hand for prototyping. But you might also want to see a platform that can run big programs (e.g., web servers) and connect big things like video projectors or televisions. Prepare to swim to the deep end, and advance to Raspberry Pi in the next chapter.

4/Sensors and the Raspberry Pi

Do you wish there was a $35 Linux computer with full HD display?

In this chapter, you will process sensor data using a Raspberry Pi, an inexpensive credit card-sized computer that runs Linux. The Pi is a great board to use as a multimedia server, a safe device to try random acts of programming on, and a fun gizmo to learn the Linux command line with.

Thanks to the Pi's general-purpose input/output (GPIO) pins, you can also wire up your own custom circuits. Some of the benefits of Raspberry Pi are:

- Cheap ($35 US)
- Connects to a full HD display
- Connects to the Internet
- Runs Linux, the free operating system
- Can run full versions of Apache web server and SSH server

Unlike Arduino, the Pi does not have dedicated pins for analog input, but you'll learn how to get around that limitation and add analog input using a special integrated circuit, the MCP3200.

You'll be able to connect all the same sensors you worked with in the Arduino chapter, but the steps to wire and program them are different.

You can start from the very beginning. To install "Hello, world," learn the command-line interface, and control an LED with Raspberry Pi, see Appendix C. Once your "Hello, world" works, return here to play with a button.

Figure 4-1. *Raspberry Pi*

Project 13: Momentary Push Button

Yes, the button strikes again! Many digital sensors behave like a button. They are on or off, 1 or 0, true or false. In this project and "Project 15: Adjustable Infrared Switch" on page 73, you'll learn to use a button and an infrared proximity sensor with Raspberry Pi.

Parts

You need the following parts for this project:

- Momentary push button
- Raspberry Pi
- Female-male jumper wires, black and green
- Breadboard

Before you build any circuit on your Raspberry Pi, you need to shut down your Raspberry Pi and disconnect it from a power source. To shut it down, click the shutdown button in the lower right of the screen, and choose the shutdown option. You can also type `sudo halt` from a terminal (LXTerminal) to begin the shutdown procedure. Wait for the shutdown process to complete before unplugging your Raspberry Pi. Double- and triple-check all connections you made before powering it up again.

Build It

Build the circuit as shown in Figure 4-2. You'll need to follow these steps:

1. Orient the breadboard as shown.
2. Insert the momentary push button anywhere on the breadboard.
3. Choose a breadboard column with a pair of button leads in it. Plug the male end of a jumper wire into that row, and connect the wire's other end to GND (pin 6 in Figure 4-2) on the Raspberry Pi.
4. Plug another jumper into a breadboard column with the other pair of button leads, and attach the wire's other end to pin 5 on the Raspberry Pi.

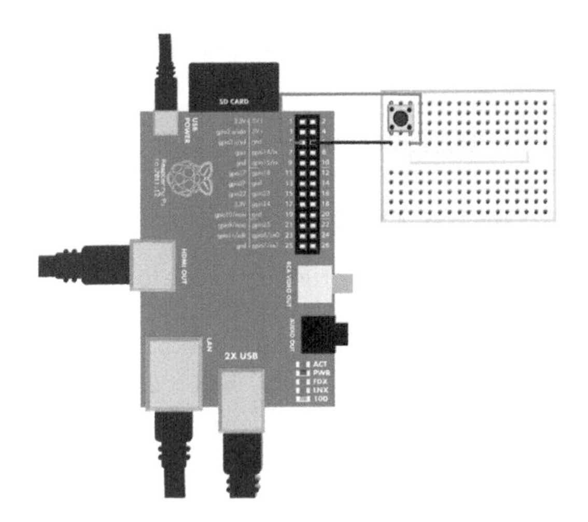

Figure 4-2. *Button on breadboard*

The button circuit has only one component: a button placed between GPIO 3 (physical pin 7) and a ground pin. When you press the button, GPIO 3 becomes connected to the ground.

On the Raspberry Pi, each GPIO pin has two numbers: one corresponding to its purpose (BCM, or Broadcom numbers), and another corresponding to its physical location. The purpose (BCM) numbers are the numbers you'll use in your programs, as shown in Table 4-1.

Table 4-1. *Pins used with button*

GPIO pin (BCM)	Board pin (place in header)
GPIO 3	5
GND	6

What happens when the button is not pressed (open)? For your program to work reliably, the pin must be connected to *something*. A pin that's not connected is *floating*, and the value from such a pin will vary with ambient electrical fields. As you might remember from "Pull-Up Resistors and Arduino" on page 37, we don't want floating pins.

There is a built-in pull-up resistor on pins GPIO 2 and GPIO 3. The internal pull-up connects the pin to +3.3 V through a 1800 Ω (1.8 kΩ) resistor. The pull-up resistor is always connected on these pins. When the button is not pressed, the pull-up resistor pulls GPIO 3 to HIGH, +3.3 V.

Be careful when connecting wires to the pins! For example, you never want to connect a GPIO to +5 V, because it could physically damage your Raspberry Pi. The pins use 3.3 V and can't handle 5 V. Before you power up your Raspberry Pi after making connections, double- and triple-check them to be sure you made them correctly.

Run the Button Code

With the button connected, you can run some code to monitor its value. Open the Terminal (LXTerminal), and follow the steps described in this section.

You should make GPIO work without root first (see Example C-1). If you want to quickly try out GPIO with root powers, use *sudo tee* instead of *tee* after the pipe. However, you really should make GPIO work as a normal user.

Export the GPIO 3 pin so that you can use it in your code (don't type the $; it indicates the shell prompt that you see in the terminal window):

```
$ echo 3|tee /sys/class/gpio/export
```

This created a new folder called *gpio3*. You can see it by typing *ls /sys/class/gpio/*.

To read a pin, you need to configure it to "in" mode:

```
$ echo in|tee /sys/class/gpio/gpio3/direction
```

Now you can read the value (don't hold down the button while you do this):

```
$ cat /sys/class/gpio/gpio3/value
1
```

Because you weren't pressing the button, the internal pull-up resistor pulled GPIO 3 to 3.3 V, HIGH, so you got the value `1`.

Next, hold down the button and read the value again (for a cool trick, you can recall the previous command by pressing the up arrow, so you don't have to type it again):

```
$ cat /sys/class/gpio/gpio3/value
0
```

Well done! You can now read values from digital sensors. Even though you'll soon practice with another digital sensor, the program will be very similar to this one.

Figure 4-3. *Buttons don't need to be small and black*

Troubleshooting

Before correcting any connections, be sure to power down your Raspberry Pi. Power it back up when you are sure the connections are correct. Here are some common troubleshooting issues:

Why do I always get a value of 0?
: Did you put your four-legged button the right way round? If you mistakenly connected GPIO 3 and ground to leads that are always connected to each other, that's exactly the same as connecting them directly. GPIO 3 would be always connected to ground (GND, 0 V), so you'd always get a value of 0.

Why do I always get a value of 1?
: This means that GPIO 3 is *never* connected to ground. A likely reason is that the button is not connected. Did you use the correct pins? Did you put all the jumpers and leads to correct rows?

Permission denied?
: These commands assume that you made your GPIO work without root, as explained in "Using GPIO Without Root" on page 114. If you insist on running everything as root (which we maintain is a dangerous, less stable way), you can test it as root by putting `sudo` in front of every *tee* com-

mand. For example, you could write *echo 3|sudo tee /sys/class/gpio/export*. But of course it's much better to avoid unnecessary use of `sudo`.

Works with sudo, but not as normal user (without sudo)?
Go through the steps in "Using GPIO Without Root" on page 114 again.

In the following projects, we'll show you how to use the GPIO pins from a Python program. All of the sensor projects in this book use Python.

Python is one of the easiest languages to learn. When we code for fun, most often it's in Python. Before we get into programming GPIOs with Python, let's get you up and running with the language.

Hello, Python World

Traditionally, the first program to write in any language is "Hello, world." It lets you test to confirm that everything is working. If you are a beginner, you'll also get assurance that you know how to write a Python program!

Whenever you start programming, you should always start with a "Hello, world"—on any language, on any platform.

You'll write your Python program just like any text file. The best place to store your program files is in your home directory (*/home/pi*). If you're not already in your home directory, you can change to it with the terminal (LXTerminal) command `cd`:

```
$ cd /home/pi
```

Start editing a new file with *nano*, a command-line text editor for editing plain-text files. Its purpose is similar to other plain-text editors such as Notepad, gedit, vi, and Emacs. Type this command:

```
$ nano hello.py
```

The file needs only one line. This is Python for you: easy things are so easy, but hard things are still possible. Add this line to the file:

```
print("Hello, world!")
```

The code has just one line. `print()` is a command, but it behaves like a function. It gets one argument, the string "Hello, world!" As you probably guessed, you are asking the computer to print the string to the terminal.

Now save your work by pressing Ctrl-X (exit). When prompted about whether you want to save, type "y" for yes. Finally, confirm the name, "hello.py", by pressing Enter or Return.

Now it's time to run your Python program. This is a big moment if this is your first step in Python. Type the command shown after the $ prompt. You'll see the output as soon as you press Enter:

```
$ python hello.py
Hello, world!
```

Your Python program printed the string "Hello, world!" Now you know you can write Python, and that your Python environment works.

Project 14: Blink an LED with Python

When starting with a new feature, it's good practice to write a "Hello, world" program. It's a program that tests the feature as simply as possible.

In this case, to start working with GPIO and Python, you'll write a program that blinks an LED. Knowing how to do this in Python, it will be much easier to use this skill as a part of your robot or gadget projects.

Parts

You need the following parts for this project:

- Raspberry Pi
- Female-male jumper wires, black and green
- An LED
- 470 Ω resistor (four-band resistor: yellow-violet-brown; five-band resistor: yellow-violet-black-black; the last band will vary depending on the resistor's tolerance)
- Breadboard

Build the LED Blink Project

Connect an LED and a resistor to GPIO 27 as shown in Figure 4-4.

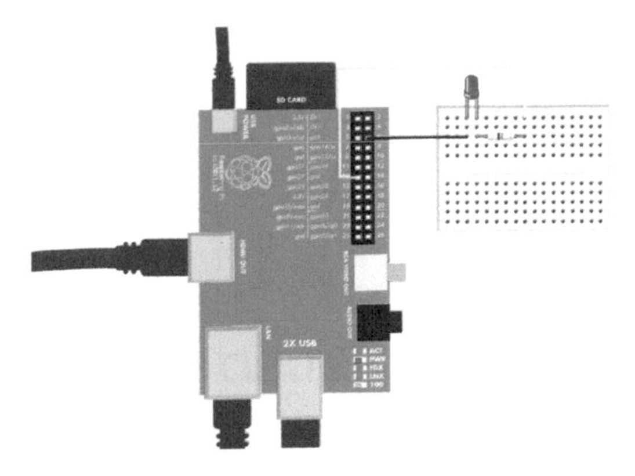

Figure 4-4. *Hello, LED*

Run the Code

Open a terminal, then change to your home directory *cd /home/pi/* if you're not already there. Create a new file with *nano led_hello.py*, type in the code from Example 4-1 (or download it from this book's website (*http://botbook.com*)), and save it with Ctrl-X, press y, and then Enter or Return.

Run the program with:

```
$ python led_hello.py
```

The LED lights up for a moment, then dims again. Did you succeed in lighting up the LED? You can now use GPIO pins with Python.

If you get "IOError: [Errno 13] Permission denied," simply try again. If the problem persists, verify that you have enabled GPIO without root using the udev rule (see "Using GPIO Without Root" on page 114).

This code shows you the secret sauce of GPIO manipulation from Python. Later on, you'll see examples that use a *library* (a file containing code that does all the work of manipulating the GPIO for you), so those code listings will be much shorter.

As you saw earlier in "Project 13: Momentary Push Button" on page 64, you need to manipulate files in the Linux filesystem to work with GPIO pins.

Want to save typing? Download the code (*http://getstarted.botbook.com*), unzip it, and copy it to */home/pi* on your Raspberry Pi SD card.

Example 4-1. Hello, LED Python code

```
# led_hello.py - blink external LED to test GPIO pins
# (c) BotBook.com - Karvinen, Karvinen, Valtokari

"led_hello.py - light a LED using Raspberry Pi GPIO"
# Copyright 2013 http://Botbook.com */


import time     # ❶
import os       # ❷

def writeFile(filename, contents):       # ❸
        with open(filename, 'w') as f:   # ❹
                f.write(contents)        # ❺

# main

print "Blinking LED on GPIO 27 once..."          # ❻

if not os.path.isfile("/sys/class/gpio/gpio27/direction"):       # ❼
        writeFile("/sys/class/gpio/export", "27")        # ❽

writeFile("/sys/class/gpio/gpio27/direction", "out")     # ❾

writeFile("/sys/class/gpio/gpio27/value", "1")  # ❿
time.sleep(2)   # seconds        # ⓫
writeFile("/sys/class/gpio/gpio27/value", "0")  # ⓬
```

❶ Import a library to give you more commands (functions) to work with. Here we are importing the time library, a very common library because it has the `time.sleep(seconds)` function that makes your program pause for a while.

❷ The OS library has commands that let you work with the underlying operating system. With `os.path.isfile(filename)`, you can test whether a file corresponding to the value in `filename` exists. Because the OS library is an abstraction of the operating system's features, the same code can work on Linux, Mac, and Windows.

❸ Define a new function called `writeFile()`. Defining a new function saves you from typing the same commands over and over again. Instead of copying all of the code in this function each time you need it, you can simply call the function `writeFile("foo.txt", "Hello")`. This function takes two parameters: a filename for the name of the file to write, and the contents to put into the file. If you called `writeFile("foo.txt", "Hello")`, the values between the parentheses will be assigned to *variables* inside the function. When `writeFile()` is executed, the variable `filename` will have the value "foo.txt" and the variable `contents` will have the value "Hello".

❹ Open `filename` for writing (*w*). This defines a new file *handle* named `f` that you can use to work with the file in later lines of code. The `with` block is the easiest built-in way to manipulate text files in Python. Python will make sure that the file handle gets closed again; either at the end of the block, or if any unexpected error occurs. Of course, your newly defined function `writeFile()` makes writing files even easier because it uses `with` for you.

❺ Using the file handle `f` defined in the previous line, we write a value of variable `contents` to the file. File handle `f` is an *object*, and `write()` is a *method* of this object.

❻ Print some text to help you debug your program. For example, if you see the text appear onscreen but the LED doesn't light, you already know that your program ran but that the problem is probably in the GPIO connections. `print()` can be used in two ways. You can call it with the parameters in parentheses as in *print("foo")*, just like any function. Or, if you're feeling lazy, you can use this shorter syntax that leaves out the parentheses.

❼ Test whether the file named *direction* already exists. If not, then execute the indented block below. This test is needed because exporting the same pin twice causes an *exception* that would make your program crash.

❽ Export the pin. Exporting creates the virtual[1] files to manipulate the pin such as *direction* and *value*. The export is done by simply writing the pin number "27" to the (virtual) text file *export*. This block is indented (there's white space to the left of the line) under the `if`, so it's only executed if the `if` condition above it evaluates to true.

❾ Write the string "out" to the text file *direction*. Here you call the function `writeFile()` that we defined earlier. When you set the direction to "out," you can turn the pin on or off.

❿ Write the string "1" to the text file *value*. This turns the GPIO 27 pin to 3.3 V (HIGH). This causes current to flow through the LED and the resistor to GND (0 V). The current lights up the LED.

⓫ Wait for two seconds. All the pins stay in the state they are in, so the LED stays on. Because `sleep` was imported from the `time` library, you have to specify the library's *namespace* (`time`) before it, as in *time.sleep()*. When your program sleeps, it takes very little computing power, and other programs running in the background can have more time.

1. So called because they don't really exist as files on stable storage (SD card).

⑫ Write "0" to value, turning off the GPIO pin. Both leads of the LED are now at the same voltage, so there is no current flowing, and the LED goes dark. Because this is the last command, the execution of the program ends.

When you are looking at a program, you should think through it in the order it is executed. First, libraries are imported to give you more functions to work (`import time...`).

Next, the first code in the *global scope* is run (*print "Blinking..."*). To simplify, it's the first command that's flush left, without any indentation (whitespace) before it.

The program execution continues line by line. The function `writeFile()` is only run when it's called (*writeFile("/sys/...")*). It's worth noting that functions are not run when they are defined (`def writeFile(filename, contents)`), even though the definition appears before the lines of code and at the same level of indentation.

Finally, the last line is executed (`writeFile("/sys/...")`). Then the program exits, and the executions stop.

Now that you know the secret sauce, the rest of the examples will use the *botbook_gpio.py* library to make the programs shorter. That library will do all of the work of manipulating files for you.

Project 15: Adjustable Infrared Switch

An adjustable infrared switch senses when something is near. You can configure the distance at which it is triggered by a nearby object. This sensor only tells you when something is near, not its actual distance.

Parts

You need the following parts for this project:

- Infrared sensor switch
- Two 1 kΩ resistors (five-band: brown-black-red-brown-any; four-band: brown-black-red-any)
- Raspberry Pi

Build the IR Switch Project

Reading any digital resistance sensor is similar to reading a button, so both the connection and the code should already look familiar to you if you've tried connecting a button to the Raspberry Pi.

Let's connect the infrared (IR) sensor switch to the Raspberry Pi. There are some differences from the connections for the button. The button just has a data pin and a ground connection. The IR sensor has some additional parts: a red +5 V pin and two resistors connected to the yellow data pin. This is because the IR sensor provides a new challenge: GPIO data pins can only tolerate 3.3 V, but the sensor works at 5 V.

You can lower the voltage of the data pin with a *voltage divider* (see "Voltage Divider" on page 76), which you can make with two resistors. A voltage divider is a circuit that lowers a voltage by dividing it between two resistors (one connected to a signal, the other to ground).

Make sure the Raspberry Pi is shut down and disconnected from power. Build the circuit as shown in Figure 4-5.

Usually, IR switches have black minus, yellow data, and red plus. We also have some models that have a green minus wire instead of a black one.

If your IR sensor switch has female pin headers, simply use jumper wire to connect it to the breadboard.

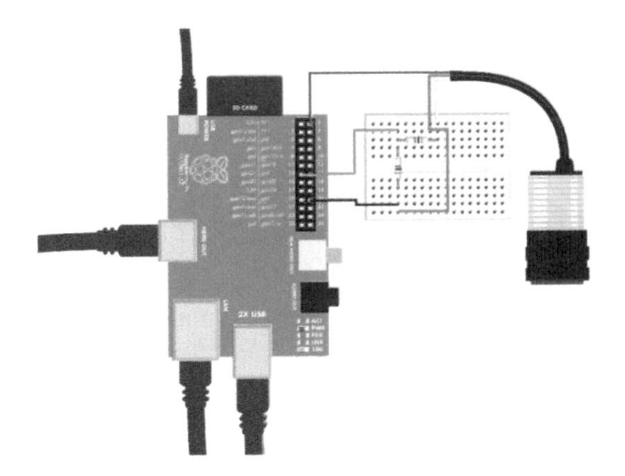

Figure 4-5. *Adjustable infrared switch connected to Raspberry Pi*

Run the Code

Power up the Raspberry Pi, then upload the code shown in Example 4-2.

Try it out:

```
$ python adjustable-infrared-sensor-switch.py
Something is inside detection range
```

Run it many times to get many measurements.

Example 4-2. Adjustable infrared switch code

```
# adjustable-infrared-sensor-switch.py - is object within predefined distance?
# (c) BotBook.com - Karvinen, Karvinen, Valtokari


import botbook_gpio as gpio     # ❶

gpio.mode(27, "in")     # ❷
x = gpio.read(27)       # ❸
if( x == 0 ):   # ❹
        print "Something is inside detection range"     # ❺
else:   # ❻
        print "There is nothing inside detection range" # ❼
```

❶ The *botbook_gpio* library contains the Python code for manipulating GPIO pins. For it to work, the *botbook_gpio.py* file must be in the same directory as this program, *adjustable-infrared-sensor-switch.py*. If you get an import error when you run it, verify that you can see both files in the same directory when you run the `ls` command. *botbook_gpio.py* simply reads and writes files under */sys/class/gpio/*, as you saw in "Project 14: Blink an LED with Python" on page 69.

❷ Setting the gpio27 pin to "in" allows its value to be read.

❸ Read the current value into new variable we call `x`. The value will be either 0 or 1.

❹ Value 0 means something has been detected. An infrared sensor switch doesn't tell the distance; it just tells whether something is nearer than the distance to which it's calibrated. An equality comparison uses two equals characters (`==`). The comparison operation, (`==`), is different from value assignment, (`=`).

❺ Print to the terminal that something was detected. This indented block under `if` only runs when the `if` condition is true; in this case, it runs if the value of variable `x` is zero.

❻ The indented block under `else` only runs when the `if` condition is not true. In this case, the one-line block under `else` runs if the value of variable `x` is not zero.

❼ Print to the terminal that nothing was detected. As this is the last instruction in the program, the execution of this program ends after this line. If you want to make another measurement, run your program again.

Voltage Divider

How can you convert voltage from 5 V to 3.3 V? As most Arduino and Arduino-compatible components use +5 V as HIGH, you will meet this question often when you try to use something with the Raspberry Pi.

The red wire supplies the IR sensor with +5 V, and the IR sensor outputs +5 V. This is much more than the Raspberry data pins can take, as their HIGH is +3.3 V. Connecting a GPIO pin to +5 V will likely destroy that pin.

When the IR sensor yellow wire is at zero, all is well because the Raspberry Pi ground (0 V, LOW) is exactly the same as the ground in the IR sensor. That's because the IR sensor ground is connected to the Raspberry Pi ground (black wire in Figure 4-5).

But what about the IR sensor HIGH state, +5 V? To protect the Raspberry Pi data pins, you must ramp down the voltage from 5 V to 3.3 V. This is easily done with a voltage divider.

The big 5 V is connected to ground (black wire, 0 V, GND) through two resistors. On the 5 V side, the voltage is obviously 5 V. Just as obviously, on the ground side of the two resistors, the voltage is 0 V. But in between, the voltage gets split according to the values of the resistors. You want 3.3 V for the Raspberry Pi:

```
3.3 V / 5 V = 0.66 = 66 %
```

So, you want the Raspberry Pi data pin to measure the voltage 66% on the way from ground to 5 V. Your total resistance is 2 kΩ (two resistors in series, 1000 Ω each):

```
66% * 2000 Ohm = 1320 Ohm
```

You won't find resistors in every possible size, so a resistor near that will probably do. The parts list mentions two 1 kΩ resistors. Putting the data pin between those resistors, you get 50% of the 5 V:

```
50% * 5V = 2.5 V
```

This is well above the minimum level for a HIGH signal, which is about half of the nominal 3.3 V HIGH level, 1.65 V. Because 1.65 V < 2.5 V, Raspberry considers 2.5 V HIGH.

But how did we come up with the 2 kΩ of resistance for the voltage divider? The exact amount doesn't matter. There must be enough resistance so that ground and 5 V don't short circuit. An air gap would achieve that, so the other requirement is that resistance must allow some electricity pass through. Thus, some kΩs is good.

You can halve the 5 V by building a voltage divider with two resistors. And now you know that lowering voltage doesn't require mysterious integrated circuits or breakout boards—just two cheap resistors.

Analog Resistance Sensors

Analog resistance sensors change their resistance according to how they respond to something in the physical world. For example, a light-dependent resistor will have low resistance in bright light, and high resistance in the dark.

The Raspberry Pi can't directly measure resistance, apart from two states: on and off. To measure resistance, you need an *analog-to-digital converter* (ADC). Unlike with the voltage divider, this will require a mysterious integrated circuit (that won't be mysterious to you in a moment).

It's a bit easier to measure analog resistance with Arduino. Because Arduino has a built-in ADC, there will be fewer wires and components in your connections. Raspberry Pi needs an external ADC chip to read analog resistance, which means more wires that could come loose or be connected incorrectly.

An advanced option for measuring analog resistance with the Raspberry Pi is using only sensors that communicate over industry standard protocols like I2C or SPI. Using I2C and SPI is out of the scope of this book, but you can learn about it in *Make: Sensors*.

Another option is to use Arduino for connecting to sensors and communicating with Raspberry Pi over USB serial. A similar setup using a computer rather than a Raspberry Pi is shown in *Make: Arduino Bots and Gadgets*. In this book, you'll work with the conceptually simple approach of reading analog resistance sensors with an ADC chip.

Project 16: Potentiometer to Measure Rotation

Turn up the volume on a radio—you're probably using a potentiometer. They are everywhere! Whenever you turn a knob to gradually change something, chances are that you are using a potentiometer. There are also hidden potentiometers (*trimmers*) inside many devices that technicians can adjust with a screwdriver.

The potentiometer (or *pot*) is the archetype of variable-resistance sensors. Once you learn to work with a pot, you can easily see similarities with other variable-resistance sensors.

This project teaches you how to read analog-resistance sensors with Raspberry Pi. Analog readings are interesting because you have to use an external analog-to-digital converter (ADC). In this project, you'll learn to use the MCP3002 ADC. It's controlled over SPI, an industry standard protocol.

Parts

You need the following parts for this project:

- Potentiometer
- MCP3002 (analog-to-digital integrated circuit)
- Raspberry Pi
- Jumper wires (male-to-female and male-to-male)
- Breadboard

The Raspberry Pi does not have female header connections for its GPIO pins. Instead, you will need to use a Pi cobbler or male-to-female jumper wires. Either option works, but the cobbler eliminates the concern of shorting out the GPIO pins by accidentally touching them together.

Build It

Build the potentiometer circuit shown in Figure 4-6.

Install SpiDev

All analog readings in this book use the MCP3002 analog-to-digital converter. The MCP3002 uses SPI, an industry standard protocol, to communicate with the Raspberry Pi. That's why all analog sensors in this book need the spidev library.

Enable Internet on your Raspberry Pi. Simply use a normal Ethernet cable to connect the Raspberry Pi to your wireless access point or switch. The Raspberry Pi receives an IP address automatically. You can test your Internet connectivity with the Midori web browser. Try *http://botbook.com*!

Install the requirements with `apt-get` (it uses the Internet, so if you get errors, check your connection):

```
$ sudo apt-get update
$ sudo apt-get -y install git python-dev
```

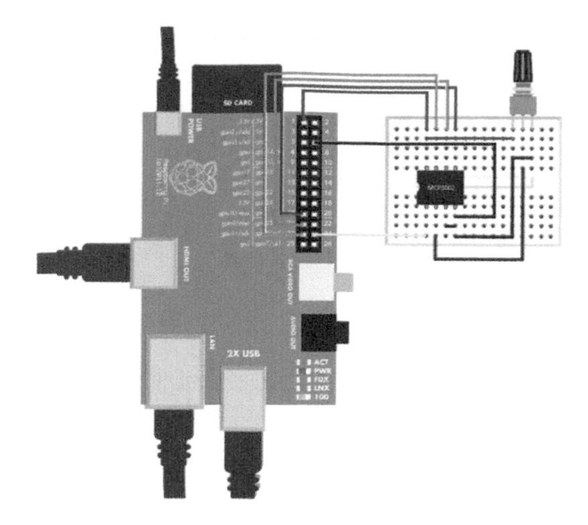

Figure 4-6. *Potentiometer connected to Raspberry Pi*

Then download and install the latest SpiDev from their version control system:

```
$ git clone https://github.com/doceme/py-spidev.git
$ cd py-spidev/
$ sudo python setup.py install
```

Allow SPI Without root

Modify udev rules, as shown in Example 4-3:

```
$ sudoedit /etc/udev/rules.d/99-spi.rules
```

Write the rule. If you edited it with nano, press Ctrl-X, then type "y," and press Enter to save.

Example 4-3. 99-spi.rules

```
# /etc/udev/rules.d/99-spi.rules - SPI without root on Raspberry Pi
# Copyright http://BotBook.com

SUBSYSTEM=="spidev", MODE="0666"
```

To avoid typos, you could alternatively copy the file with *sudo cp 99-spi.rules/ etc/udev/rules.d/*.

Finally, edit the blacklist to enable the SPI device:

```
$ sudoedit /etc/modprobe.d/raspi-blacklist.conf
```

Delete the line that says "blacklist spi-bcm2708," then press Ctrl-X to exit, type "y" then press Enter to save the file.

Reboot your Raspberry Pi. Verify that you can see some files with:

```
$ ls -l /dev/spi*
```

Can you see at least two files? Well done, now you are ready to read analog resistance sensors.

Run the Code

Once you get the circuit wired up and SpiDev installed, reading the sensor is easy with the MCP3002 library. You can download it from the website for *Make: Arduino Bots and Gadgets* (*http://botbook.com*).

Put *pot_once.py* (see Example 4-4) and the *MCP3002.py* library into the same directory. You can easily create the *pot_once.py* file with the *nano* editor. The *MCP3002.py* library is long, so it's best to download it as well (*http://botbook.com*).

Run the code:

```
$ python pot_once.py
```

View your result. Turn the knob of the potentiometer, and run the program again. In Linux, you can press the up arrow to retrieve an earlier command so you don't need to keep typing it over and over.

Example 4-4. Reading a potentiometer

```
# pot_once.py - measure resistance of a potentiometer, once
# (c) BotBook.com - Karvinen, Karvinen, Valtokari

import botbook_mcp3002 as mcp   # ❶

x = mcp.readAnalog(0,0) # ❷
print(x)        # ❸
```

❶ Import the MCP3002 library. The file *MCP3002.py* must be in the same directory as this program (*pot_once.py*). All the heavy lifting, like the SPI interface, is done with this library.

❷ Read the analog value from the MCP3002 chip. The `readAnalog()` function is from the MCP3002 library, as you can see from the namespace prefix "MCP3002.". The parameters device (0) and channel (0) correspond to the circuit you built. The `readAnalog()` function returns a number between 0 and 1023. This number is saved to a new variable called `x`.

❸ Print the value of variable `x`.

Troubleshooting? If running *pot_once.py* gives you this error...

```
spi.open(0, device) .. IOError: [Errno 2] No such file or directory
```

...verify that you modified */etc/modprobe.d/raspi-blacklist.conf* as instructed in "Allow SPI Without root" on page 79, then rebooted. You should see some files if you run the command **`ls /dev/spi`**.

Repeat Readings

Most embedded devices have a never-ending loop that keeps running forever (Arduino has this: the `loop()` function runs until you shut the device down). The program will do its thing, wait for a short while, then do it again. It makes sense that an embedded device works as long as there is power: you wouldn't want your car's assisted steering or your office air conditioning to stop and require restarting all the time.

Example 4-5 shows how this is done in Python. Keep the same circuit as before, but run the new code. Now you can continuously get new readings while you adjust the potentiometer. Press Ctrl-C to kill the program and get your command prompt back.

Example 4-5. Reading a pot over and over

```
# pot_repeat.py - continuously measure resistance of a potentiometer
# (c) BotBook.com - Karvinen, Karvinen, Valtokari


import botbook_mcp3002 as mcp   # ❶
import time     # ❷

while(True):    # ❸
        x = mcp.readAnalog(0,0) # ❹
        print(x)        # ❺
        time.sleep(0.5) # seconds       # ❻
```

❶ Import the library for the analog-to-digital converter chip. Just as before, *MCP3002.py* must be in the same directory as this program, *pot_repeat.py*.

❷ Import the library that supplies the `time.sleep()` function.

❸ Repeat the block below as long as the condition is true: forever. Press Ctrl-C to kill the program.

❹ Read the analog value just like you did in *pot_once.py*.

❺ Print the value of `x`, just like in *pot_once.py*.

❻ Wait for a while. Whenever there is an infinite loop, you must make sure you give time to the operating system and other programs. Otherwise, your program could take 100% of CPU power of any one core, even one running a zillion gigahertz.

Why Does a Potentiometer Have Three Legs?

You may have noticed that you didn't use a pull-up, pull-down, or voltage divider, and you probably also noticed that a potentiometer has three leads.

One side is connected to ground (LOW, 0 V, black wire, GND, "negative"). The other side is connected to positive (HIGH, 3.3 V, +, red wire). The middle lead is connected to the data pin.

Turning the knob, you can select where the middle lead touches the resistor. Because there is a resistor on either side of the pot (between the middle lead and GND, and between the middle lead and +V), the pot itself is a voltage divider!

If the knob is turned to the minimum setting, there is nearly no resistance between the data pin (middle) and GND. All the resistance is between the data pin and +3.3 V. So the data pin gets 0 V, and `MCP3002.readAnalog(0,0)` returns 0.

When you turn the knob to the maximum, there is no resistance between the data pin and +3.3 V. All the resistance is between data and ground. So the data pin gets 3.3 V, and `MCP3002.readAnalog(0,0)` returns 1023.

When the knob is anywhere between maximum and minimum, you get a voltage between LOW (0 V) and HIGH (3.3 V). `MCP3002.readAnalog(0,0)` returns a number between 0 and 1023.

The voltage seen by the data pin is proportional to the position of the pot.

Project 17: Photoresistor

Resistor by night, wire by day. A photoresistor reduces its resistance when it's bright. It's also known as a light-dependent resistor (LDR).

The LDR changes resistance. From the Raspberry Pi's point of view, isn't this the same idea as a potentiometer?

In fact, the photoresistor can use the same code as a potentiometer. The only difference is the number of leads on the sensor. As you can see from Figure 4-7, an extra resistor is used to form a voltage divider. You may remember that a voltage divider requires two resistors; the photoresistor is the second (and variable) resistor in the divider.

Parts

You need the following parts for this project:

- Photoresistor (10 kΩ suggested)
- Resistor (10 kΩ, or choose a value that matches the photoresistor)
- MCP3002

- Raspberry Pi
- Jumper wires (male-to-female and male-to-male)
- Breadboard

The 10 kΩ resistor coloring is brown-black-orange-any (four-band) or brown-black-black-red-any (five-band).

Build It

Wire up the circuit as shown in Figure 4-7.

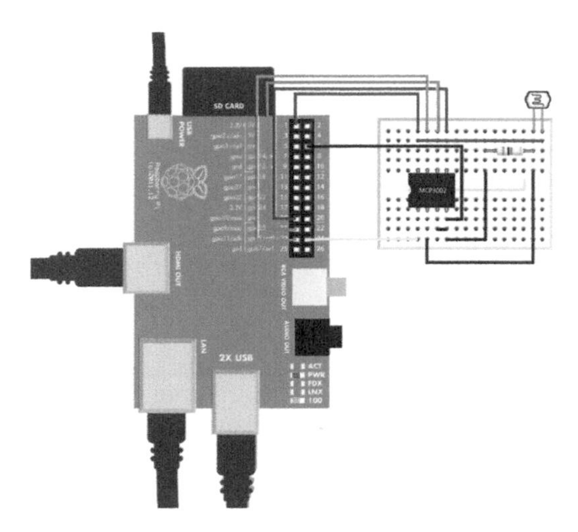

Figure 4-7. *Photoresistor connected to Raspberry Pi*

Run the Code

After you've built the circuit, run the pot code again:

```
$ python pot_repeat.py
```

When you see numbers appearing, try changing the value. Cover the LDR with your finger, and you see low numbers. Point a bright light at it, and the numbers increase.

Once you are done, kill the program with Ctrl-C.

Playing with Resistance Numbers

If that felt too easy, let's try playing with the measured numbers. There are three common ways to look at the measured resistance:

- Raw number, from 0 to 1023
- Voltage, from 0 V to 3.3 V
- Percentage, from 0% to 100%

You have already read the raw value with a potentiometer and light-dependent resistor. It's the value you get from this line of code:

```
MCP3002.readAnalog(0,0)
```

The raw number doesn't have an obvious unit—it's just an integer (a whole number) from 0 to 1023. This is the number you always start with when you take a sensor measurement.

It's easy to convert a raw value to a percentage, from 0% to 100%. The minimum raw value 0 is 0%. The maximum value 1023 is 100%. Any value x in-between is x/1023. For example, the raw value 348 is:

```
348/1023 = 0.34 = 34%
```

So the raw value 348 is 34% of the maximum value.

```
raw = MCP3002.readAnalog(0,0)
percent = raw/1023.0    # ❶
```

❶ To get a floating-point result, you must divide by a floating-point number. In Python (and many other languages), dividing by an integer (a whole number) gives you a counter-intuitive whole number result. In Python, 1/2 == 0, but 1/2.0 == 0.5. Notice how 2 is an integer and 2.0 is a floating-point number.

You might already know that a percent is a one hundredth part. This means that 0.34 is exactly 34%. When you speak about parts, percentages are practical. When programming, a floating-point number 0.34 is practical, as you can easily multiply a number with this.

In your own projects, you measure something to change something. You might use a potentiometer to set the volume of a beep. Or you could use an LDR to turn lights brighter when it's dark outside. In these cases, it's convenient to multiply by a percentage.

To get the voltage of the data pin, you can multiply by the percentage. The maximum voltage (HIGH) of a Raspberry Pi pin is 3.3 V (100%) and the minimum (LOW) is 0 V (0%):

```
raw = MCP3002.readAnalog(0,0)
percent = raw/1023.0
voltage = percent * 3.3
```

Project 18: FlexiForce

The FlexiForce sensor (see Figure 3-14) measures pressure. The harder you squeeze the round end of the sensor, the more electricity it lets through. Just as with a pot and LDR, you can use the same code.

Parts

You need the following parts for this project:

- Flexiforce (11 kg, 25 pound version recommended)
- 1 MΩ resistor (five-band: brown-black-black-yellow-any; four-band: brown-black-green-any)
- MCP3002
- Raspberry Pi
- Jumper wires (male-to-female and male-to-male)
- Breadboard

Build It

Build the FlexiForce circuit shown in Figure 4-8. As the FlexiForce has just two leads; this connection is similar to the light-dependent resistor (LDR). There is a second resistor to form a voltage divider

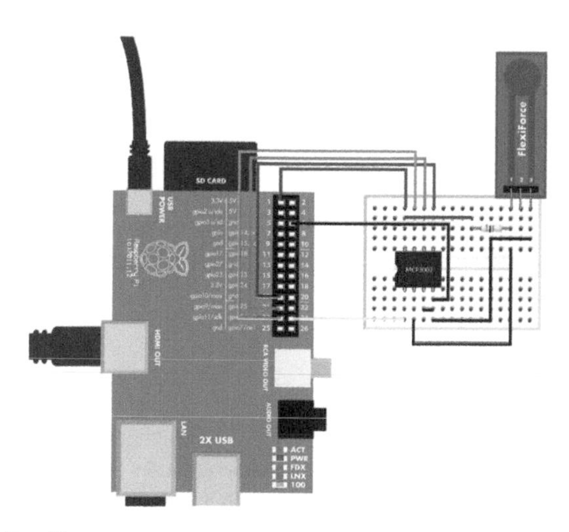

Figure 4-8. *FlexiForce squeeze sensor connected to Raspberry Pi*

What about the third pin? The FlexiForce has a third pin in the middle, but it's not connected to anything. It's just there to make it easier to connect the FlexiForce to a breadboard. The outer pins are live.

Run the Code

Run the potentiometer code again:

```
$ python pot_repeat.py
```

Play with the values. Try leaving the sensor untouched. Then squeeze the round end hard.

You've tried three analog resistance sensors: a pot, an LDR, and the FlexiForce. Do you recognize a pattern? All of these sensors can use the exact same code. The circuit with the analog-to-digital converter (ADC, MCP3002) is nearly identical for the different components. The only difference is the second resistor for the sensors (the first resistor in the voltage divider) that have just two leads.

Project 19: Temperature Measurements (LM35)

The LM35 reports temperature by changing its resistance. In its default configuration, the LM35 can measure temperature from 2 °C to 150 °C.

Parts

You need the following parts for this project:

- LM35 temperature sensor
- MCP3002
- Raspberry Pi
- Jumper wires (male-to-female and male-to-male)
- Breadboard

Build It

Build the circuit shown in Figure 4-9. Wires that are always in 3.3 V (same as data pin HIGH) are red in the diagram. The dangerous (to Raspberry Pi's pins) +5 V wire is yellow. Don't connect the yellow +5 V wire to data pins.

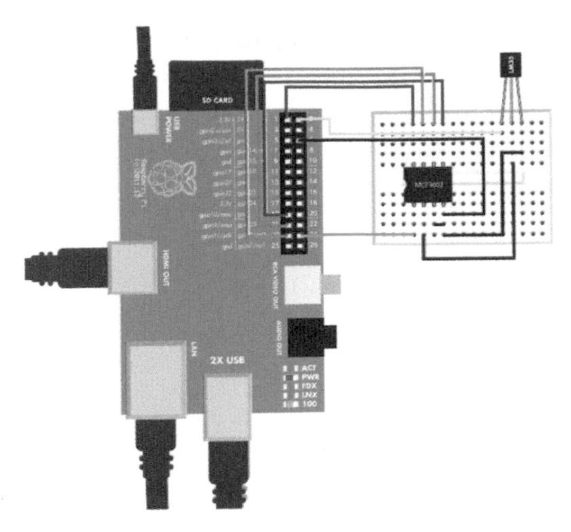

Figure 4-9. *LM35 temperature sensors connected to Raspberry Pi*

As LM35 has three leads, the connection is similar to the pot circuit. The only difference is the 5 V voltage required by the LM35.

The LM35 uses +5 V, enough to damage the Raspberry Pi's microcontroller IC. Mistakenly connecting 5 V to a data pin, even for a moment, damages the Raspberry Pi. Most likely, the damage is permanent, destroying some or all of the data pins in the GPIO header. That's why you use different colors for safe +3.3 V (red) and dangerous +5 V (yellow). Be careful where you connect +5 V. Just be careful to make correct connections and you'll be fine.

Run the Code

Every analog resistance sensor thus far (pot, LDR, and FlexiForce) has worked with the same code. You can also try pot code here.

It does work, but the raw numbers are not very useful. What good does it do to know that the temperature results in a raw value such as 815?

As you saw earlier, you can convert raw values to percentages and voltages.

All components have a datasheet that describes how they are used. Datasheets can be terse and technical, which makes books like this quite useful. You can find the datasheet for the LM35 by searching the Web for "lm35 datasheet."

The datasheet for the LM35 tells us that the output voltage changes by 0.010 V per 1 °C. The inverse of this is that a 1 V change represents 100 °C. Some sample voltage-temperature pairs are:

- 0 °C, 0.0 V (this circuit doesn't measure temperatures this cold)
- 2 °C, 0.02 V (minimum measured temperature, a cold fridge)
- 10 °C, 0.1 V
- 20 °C, 0.2 V (summer in Helsinki)
- 30 °C, 0.3 V (summer in Spain)
- 40 °C, 0.4 V
- 80 °C, 0.8 V (a sauna)
- 100 °C, 1.0 V (water boils)
- 150 °C, 1.5 V (maximum measured temperature)

This also explains why we're not worried about getting too much voltage to the Raspberry Pi data pin: to get over 3.3 V, you would need 330 °C of heat.

You must convert the raw value to voltage, and voltage to Celsius. For this, let's create a new program.

Run the code shown in Example 4-6.

Example 4-6. Reading the LM35 sensor in Python

```
# lm35.py - print temperature in Celsius
# (c) BotBook.com - Karvinen, Karvinen, Valtokari


import time     # ❶
import botbook_mcp3002 as mcp   # ❷

def readTemperature():  # ❸
        data = mcp.readAnalog(0,0)      # ❹
        percent = data / 1023.0 # ❺
        volts = percent * 3.3   # ❻
        celcius = 100.0 * volts # ❼
        return celcius  # ❽

while True:     # ❾
        t = readTemperature()   # ❿
        print("Current temperature is %i C " % t)       # ⓫
        time.sleep(0.5) # seconds       # ⓬
```

❶ The `time` library is needed for `time.sleep()` in the loop.

❷ As with previous examples, the file *MCP3002.py* must be in the same directory with this program (*lm35.py*).

❸ `readTemperature()` is a function, so it's easy to add it to your own projects.

❹ Read the raw value, from 0 to 1023. This is the same command you used with the potentiometer, LDR, and FlexiForce.

❺ Convert to a percentage of maximum value, from 0.0 (0%) to 1.0 (100%). The divider must be a floating point to get a floating-point result.

❻ Voltage is calculated from the maximum voltage.

❼ Convert voltage to Celsius, using the 0.010 V per 1 °C formula from the datasheet.

❽ Return the temperature.

❾ Keep measuring and printing temperatures until the user presses Ctrl-C to kill the program.

❿ This is all you need to do to read temperature in your own programs. Declare a new variable `t`, read the temperature, and store the value there.

⓫ Print the temperature with some helpful text. `print` uses a format string, where `%i` is replaced with an integer value, the value of the variable `t`.

⓬ Wait for a while. This allows you to read the printed temperature, and also prevents the program from taking all the CPU time.

Project 20: Ultrasonic Distance

An ultrasonic distance sensor gives you the distance to a target. For example, it can tell you there is a target 58 cm from the sensor. It's one of the most popular sensors in the courses we teach.

HC-SR04 is a cheap ultrasonic sensor, costing just a couple of dollars. This is a very nice price compared to alternatives costing much more.

Parts

You need the following parts for this project:

- HC-SR04 sensor
- Two 10 KΩ resistors (four-band: brown-black-orange-any, five-band: brown-black-black-red-any)
- Raspberry Pi
- Jumper wires (male-to-female and male-to-male)
- Breadboard

Build It

Build the circuit as shown in Figure 4-10. The sensor uses a protocol that's based on pulse length, where pulses consist of changes in voltage level from HIGH or LOW. As no analog-to-digital converter is needed, the circuit is quite simple. Be careful with the +5 V pin as connecting that to any data pin will likely break the whole GPIO header beyond repair.

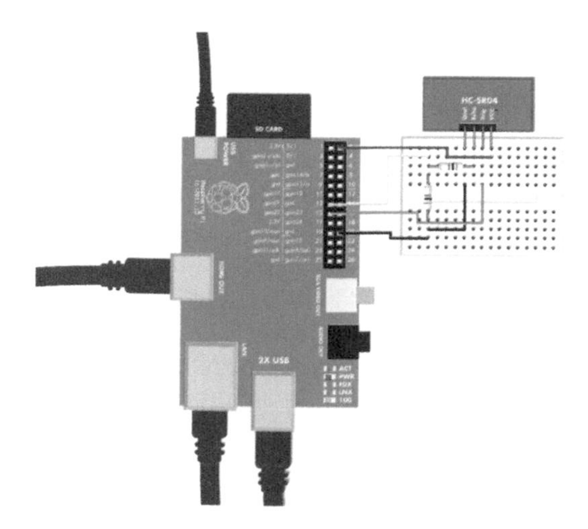

Figure 4-10. *Cheap ultrasonic distance sensor HC-SR04 connected to Raspberry Pi*

Run the Code

Run the code shown in Example 4-7.

Example 4-7. Reading the HC-SR04 in Python

```
# hc-sr04.py - measure distance with ultrasound
# (c) BotBook.com - Karvinen, Karvinen, Valtokari


import time     # ❶
import botbook_gpio as gpio     # ❷

def readDistanceCm():   # ❸
        triggerPin = 22 # ❹
        echoPin = 27    # ❺

        v=(331.5+0.6*20) # m/s  # ❻

        gpio.mode(triggerPin,"out")      # ❼

        gpio.mode(echoPin,"in") # ❽
```

```
    gpio.interruptMode(echoPin, "both")     # ❾

    gpio.write(triggerPin, 0)       # ❿
    time.sleep(0.5) # ⓫

    gpio.write(triggerPin, 1)       # ⓬
    time.sleep(1/1000.0/1000.0)     # ⓭
    gpio.write(triggerPin, 0)       # ⓮

    t = gpio.pulseInHigh(echoPin) # s       # ⓯

    d = t*v # ⓰
    d = d/2 # ⓱
    return d*100 # cm       # ⓲

dist = readDistanceCm() # ⓳
print("Distance is %i cm" % dist)       # ⓴
```

❶ If you later improve the program measure distance continuously, you'll need `time.sleep()` in the loop.

❷ The *botbook_gpio* library controls the GPIO pins. You must have *botbook_gpio.py* in the same directory as this program, *hc-sr04.py*.

❸ The `readDistanceCm()` function is defined here. This function does all the hard work. You don't need to think how it works when you call it from your program.

❹ Define a constant for GPIO pin 22, which is connected to the HC-SR04 trigger pin. The word *trigger* was chosen because it's marked "Trig" on the sensor. Whenever `triggerPin` is mentioned in this function, it will use the number 22 in that place.

❺ The GPIO pin 27 is connected to HC-SR04 "Echo."

❻ Calculate the speed of sound, which is about 340 meters per second. The number 20 is the temperature in Celsius; feel free to change it if you happen to be in a warmer or colder place. The Raspberry Pi will do a simple calculation instantly, so there is no need to calculate it with pen and paper and put what amounts to a "magic number" into code.

❼ Set the trigger pin to "out" mode, so you can control it. The trigger pin is used for sending the sound pulse with the ultrasonic speaker. This will automatically export the pin, creating the needed *direction* and *value* files.

❽ Set the echo pin to "in" so that you can read the pulse.

❾ Enable the interrupt on `echoPin`, so that you can later measure the pulse duration with `pulseInHigh()`. The interrupt will happen "both" for rising edge (start of pulse) and falling edge (end of pulse).

❿ Make sure the trigger pin is LOW, so that you can send a short pulse later.

⓫ Wait for the pin to stabilize, so that there is a clear start for the pulse.

⓬ Bring GPIO 22 HIGH (3.3 V), starting the pulse.

⓭ Wait for a very short time. GPIO 22 stays HIGH. This wait defines the length of the pulse. One millionth of a second is one microsecond. Millionth is written as a calculation, because it's annoying to try to count the zeroes in 0.000001. To get a floating-point (decimal) number, the dividers must be floating-point numbers (1000.0 instead of integer 1000).

⓮ Turn GPIO 22 off (LOW), ending the pulse.

⓯ Measure the length of the pulse coming to GPIO 27. Save the pulse duration to a new variable, `t`. The pulse length measurement is not as precise as in Arduino (see "Real Time or Fast?" on page 92). The uncertainty of the pulse measurement limits the precision of the distance measurement. Still, `pulseInHigh()` and interrupts are much better than checking the pin state with a loop.

⓰ Calculate distance `d`, using velocity `v` and time `t`. For example, if you drive one hour at 100 km/h, you move a hundred kilometers.

⓱ One way trip is half of a two-way trip. The sound goes to target and echoes back, so the distance is half of that round trip.

⓲ Convert meters to centimeters.

⓳ Measure the distance in centimeters, and save it to new variable `d`. This is how you can measure distance in your own programs. This is the first line executed after library imports.

⓴ Print some helpful text and the distance. To get rid of extra decimals, `%i` converts distance to an integer.

Real Time or Fast?

If you tried HC-SR04 with both Arduino and Raspberry Pi, you probably noticed that Arduino gives more reliable and precise results. Arduino is more *real time* than Raspberry Pi.

Real time means that operations take the same, predictable amount of time. Real time doesn't mean fast. Raspberry Pi is much faster (more calculations per second), but it's not real time.

Arduino doesn't have an operating system, so the only thing happening on its CPU is the code you've written. Raspberry Pi has a whole operating system (Linux kernel, GNU tools, daemons, etc.). At any moment, many programs are running on a Raspberry Pi. They will affect how fast your program runs, and this makes things less predictable.

You have now gone through a long journey. You can sense the world using many different approaches. The most basic, even if tedious, way is to build a circuit component by component. Usually, the easiest way is to use Arduino. And if you need a real computer, there is Raspberry Pi.

Now it's time to make your skills serve your imagination.

Build your own ideas into prototypes. Make your own robots. Innovate devices and gadgets that have not been made before. We've written several books that can help you.

To learn advanced sensors, read *Make: Sensors*. To add an EEG sensor to your repertoire, build the bot in *Make a Mind-Controlled Arduino Robot*. *Make: Arduino Bots and Gadgets* brings you six fun projects and shows how to achieve impressive results quickly.

Go make your projects!

A/Troubleshooting Tactics

No one gets everything right the first time, so it's a good idea to be deliberate in how you approach troubleshooting. Here are some tactics we use when our projects do not behave as expected:

Verify hardware with "Hello, world"
: By this, we mean get the simplest possible program working on your hardware before you introduce complexity. Traditionally, a "Hello, world" program prints a string of text onto a screen, but of course, many sensor projects don't call for a display. Our Arduino and Raspberry Pi "Hello, world" analog is usually just a short program that will blink an onboard LED on and off at a specific intervals.

Always build and test in small steps
: Racing through a project only to find it does not work is never a good experience. Our approach is to deliberately build and test as the project progresses, not wait until the end to run the first test. This tactic means we don't end up with an extremely complex problem to troubleshoot when the build is done. Testing is time well spent! Additionally, if you do create a habit of regular testing, it is much easier to go back to a known working state. Don't try to keep pushing on when things don't work. If you do, you'll end up frustrated with a pile of messy code and wiring that nobody, including you, can understand. Building in small steps does not mean that you have to understand every line of the code or inner workings of your sensor. One working component and a basic understanding of how it works is good enough.

Isolate the problem
: Let's say you are trying to make an Arduino project where a servo is controlled by moving a rotary knob (a potentiometer). Taking our advice, you built the project methodically and in small, manageable steps. But the servo is simply not moving as you expect when the knob is turned. Here is how we'd approach the issue. First, only connect the servo to your Arduino and upload the most basic code that controls servo movement. If the servo moves as expected, then it's on to test the knob in the same way you just did for the servo. If the servo does not move as expected, you've isolated the problem. If both parts work independently,

then you can be fairly confident the issue is in the code connecting the two components.

RTFM (read the friendly manual)
: Reading documentation is often overlooked as a troubleshooting resource, but it's exactly where you should look for help! If there was no manual accompanying a sensor you bought, try checking the manufacturer's website. Another great place is online forums, but you should be careful not to blindly assume that, because it's in a forum or online, it conveys the correct answer. There are lots of flawed instructions and even nonfunctional sample programs, so it's good to have some skepticism while reading. (Another thing to be aware of: the F in RTFM isn't always considered to stand for *Friendly*. It's best not to use the acronym in an email to your best client.)

Document your work
: Comment your code! No need to comment every line, but comment the code where it's tricky or where it took you a while to get it working properly. Photograph the stages of your build. It's great to have images for future use in a tutorial, but it's helpful to have them even as you are iterating through different project designs. Otherwise, you'll forget what you did in no time. If you want to kick it up a notch, publish your results online!

B/Arduino IDE Setup

Here's how to set up the Arduino integrated development environment (IDE) on various platforms.

Ubuntu Linux

Connect Arduino to your computer with a USB cable. Arduino draws power directly from the USB, so no external power supply is needed. Start the terminal application.

You can start the command-line terminal in many ways. You can open it in the main menu with Applications → Accessories → Terminal (on Xubuntu and other XFCE-based distributions such as Debian with XFCE). Super-T, also known as the *ugly key* or *Windows key*, works on many desktops. If you are using Unity in the standard Ubuntu distribution, search for "Terminal" in Dash (top-left corner).

To install the Arduino IDE, install the `arduino` package. Here's how you'd do it on Ubuntu Linux:

```
$ sudo apt-get update
$ sudo apt-get -y install arduino
```

Give yourself permission to access the serial-over-USB port (this is required by the Arduino development environment to function). The first command adds you to the dialout group, and the second command switches you into that group without you needing to log out and back in again:

```
$ sudo adduser $(whoami) dialout
$ newgrp dialout
```

Start Arduino:

```
$ arduino
```

The Arduino IDE opens.

After you have logged out and back in, you can also start Arduino IDE from the menus.

Now you're ready to test your installation. See "Hello, World" on page 99.

Windows 7 and 8

Download the latest version of the Arduino software (*http://bit.ly/arduino-dl*) and unzip the file you downloaded to any location that you find suitable (your *Desktop* or *Downloads* directory, for example).

Connect your Arduino Uno to your computer with a USB cable. Arduino draws power directly from the USB, so no external power supply is needed. Windows will start an automatic installation process for the Arduino drivers. It may fail after a while and display an error dialog.

If Windows fails to install the driver:

1. Open Windows Explorer, right-click Computer, and choose Manage.
2. From Computer Management, choose Device Manager on the left. Locate Arduino Uno in the device list, right-click it, and choose Update Driver Software.
3. Choose "Browse my computer for driver software." Navigate to the Arduino folder you extracted, open the *drivers* directory, choose *arduino.inf*, and click Next.
4. Windows will now install the driver.

Launch the Arduino IDE by double-clicking the Arduino icon inside the folder you unzipped.

Time to test your installation; see "Hello, World" on page 99.

OS X

Download the latest version of the Arduino software (*http://bit.ly/arduino-dl*) and unzip the file you downloaded. Copy it to your */Applications* folder.

Connect your Arduino Uno to your computer with a USB cable. Arduino draws power directly from the USB, so no external power supply is needed. You don't need to install a driver for OS X.

Launch the Arduino IDE by double-clicking the Arduino icon in the */Applications* folder.

Time to test your installation; see "Hello, World" on page 99.

Hello, World

Now that you have the Arduino IDE open, you can run the Arduino equivalent of "Hello, world."

First, confirm that you have the correct board selected. The Arduino Uno is the default. If you have another board, such as a Mega or a Leonardo, choose it from the Tools → Board menu.

Now you need to load the Blink test program. Choose File → Examples → 1.Basics → Blink. Click the Upload button (or choose File → Upload) to compile and upload your program to Arduino.

The first time you do this, Arduino may display an error popup: "Serial port COM1 not found." That's because you haven't chosen which serial port to use (the connection between your computer and Arduino is represented as a USB serial port). Select your serial port from the drop-down menu. On Linux, it's probably */dev/ttyACM0*; on the Mac, it may be something like */dev/usbmodem1234*; and on Windows, it's one of the COM ports.

If you see a different error message instead of a request to choose a serial port, choose your serial port from Tools → Port. If you can't figure out which port Arduino is connected to, pay attention to the ports listed, unplug the Arduino, and make a note as to which port went away. That's the Arduino port. OS X lists each port twice (e.g., as */dev/cu.usbmodem1234* and */dev/tty.usbmodem1234*). Either one will work.

While the program is uploading, Arduino's TX and RX (transmit and receive) lights blink rapidly. Finally, when the program is running, the tiny light labeled "L" is blinking.

The L LED blinking means that everything was successfully installed, and you just got your first sketch running.

Congratulations! Remember this simple procedure: if you ever get so stuck you are wondering whether Arduino is even running your code at all, return to this "Hello, world" example. Whenever you start a new program, start with a "Hello, world" to make sure that everything is working.

C/Setting Up Raspberry Pi

To make Raspberry Pi work, you need to install an operating system. Without it, you won't even get a picture on the screen.

To install an operating system, you'll need to do the following:

- Prepare a memory card: format it and extract a compressed zip archive.
- Connect cables and boot to the installer; Raspberry Pi boots automatically when you connect the power cable.
- Install Raspbian; just select it from the menu.
- Configure your Linux; it's easy using the menu that displays automatically.

Parts

You need the following parts for this project:

- Raspberry Pi Model B (the one with wired Internet)
- Micro USB cable (for power)
- 4 GB SD card (can be 8 GB too; Raspberry Pi Model B+ uses a microSD card)
- Display with HDMI port (your TV could qualify)
- HDMI cable
- USB mouse
- USB keyboard
- An SD card reader (for your regular computer; with the Raspberry Pi Model B+, you may need a microSD to SD adapter to read the card on your computer)

Optionally, it's convenient to have an Ethernet cable that connects the Raspberry Pi to the Internet.

Set It Up

The following sections describe how to set up the Raspberry Pi.

Prepare the Memory Card

To prepare the memory card, you download *NOOBS*.zip* and extract its contents to an SD card.

Did you buy an SD card with preinstalled Raspbian? You can skip this "Prepare Memory Card" section.

The SD memory card in the Raspberry Pi is the place to store data permanently, over reboots. As Raspberry Pi doesn't have a hard disk, the memory card serves the same purpose.

Some shops also sell memory cards that have the operating system preinstalled. If you have one, you don't have to install an operating system, and you can skip to the point where Raspberry Pi is booted from this card.

Buying the SD card is quite simple. The name SD defines both the technology and the physical size of the card. You want the big SD size (3.2 cm by 2.4 cm), not the smaller mini SD or tiny micro SD. Most cards nowadays are preformatted, and that's convenient.

Open a web browser on your regular computer. Go to the official Raspberry Pi website (*http://bit.ly/rasp-dl*) and download the Quick Start Guide and the latest *NOOBS*.zip* using the "Direct download" or "Download ZIP" links. The name of the *NOOBS*.zip* varies, so we've marked the version number with a star. For example, the actual name could be *NOOBS_v1_2.zip*.

If you're lucky, your memory card is preformatted when you buy it. If not, format it to FAT according to the instructions in the Quick Start Guide you just downloaded.

Put the SD memory card in the SD card reader of your regular computer. Your laptop might have an SD card reader by default. For most desktops, you need to buy a cheap external USB-connected reader.

Extract the contents of *NOOBS*.zip* to the memory card. In most modern Linux, Windows, and Mac environments, just double-click or right-click *NOOBS*.zip*. If you have trouble extracting the file in older Windows versions, download and install 7zip and extract with that.

After extraction, you should have the contents on the top level of the memory card. For example, *bootcode.bin* must be on the top level of the card and not inside a folder.

Connect Cables and Boot to Installer

Connect the cables to Raspberry Pi; they only fit in the correct holes. Connect the USB keyboard and mouse. Connect an HDMI cable between the Raspberry Pi and your television or monitor.

Start your television or monitor. Make sure it's set up to use the HDMI input to which the Raspberry Pi is connected.

Insert the SD card. The SD card only fits the slot the correct way. The SD card slot is on the short sides of the Raspberry Pi, the opposite side from the USB and Ethernet ports.

Finally, connect the micro USB cable between the Raspberry Pi and a power source. Raspberry Pi requires about 1 watt (1 W) of power, so a suitable power source is a desktop computer or a powerful USB charger.

A red PWR (power) LED lights up. Your monitor should show an image of a raspberry. This picture on the display means that Raspberry Pi is successfully running something from the SD card. Raspberry Pi doesn't have a BIOS similar to a workstation, so without software from the SD card, the screen is black.

After a moment, you see a menu of installable operating systems.

Install Raspbian

The installer menu shows you many different Linux distributions that you could install. Select Raspbian [RECOMMENDED] (see Figure C-1) and press Enter. (Or in the newest version, select "Install OS.") It's the most popular one. If you have ever used Ubuntu, Debian, or Mint, you will soon feel at home with Raspbian.

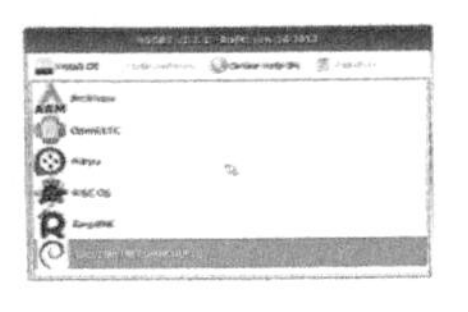

Figure C-1. *NOOBS boot screen*

Answer yes to the confirmation "All existing data on the SD card will be overwritten" by pressing Enter again. Enjoy the slideshow while Raspbian installs.

The installation takes less than 10 minutes. When installation completes, you get a pop-up "Image applied successfully." Press Enter for OK.

The Raspberry Pi reboots. You get a screen with a raspberry at the top left, and the log filling most of the screen. The log explains what's happening when Raspberry Pi boots. This can be quite entertaining to watch. Once the boot completes, you are in the "Raspberry Pi Software Configuration Tool," *raspi-config*.

Configuration with raspi-config

Raspi-config is a menu with the most common configuration tasks. You can move around with the arrow keys and Tab. To select items, press Enter or Return.

Change your password
: Move the selection to "2 Change User Password." The default username is "pi." The default password "raspberry" could be a security risk because everyone knows it, so the first task in the new system is to change the password. Choose a good password and type it at the prompt. What you type is not echoed on the screen, so you'll have to type it blindly. After typing the password twice, you get a confirmation of your successful password change. Then you return to the raspi-config menu.

Enable boot to desktop
: Even though Raspberry Pi can't replace a real workstation or a laptop, it's helpful in the beginning to have a web browser and multiple windows when working. Select "3 Enable Boot to Desktop." There is just one prompt, "Should we boot straight to desktop?" Press Enter to answer yes. Then you return to the raspi-config menu.

  Press Tab and the arrow keys to select <Finish> and Enter to choose it. In most keyboards, Tab is on the left side of the keyboard, above the Caps Lock key.

You can also configure your time zone, locale (you'll need to set this if you use a non-English language), keyboard, and more.

The big moment: "Would you like to reboot now?" Press Enter for <Yes>. Enjoy the log messages while Raspberry Pi shuts down.

Boot to Desktop

When Raspberry Pi boots, you briefly see four pixels filling the screen. Then you'll see the familiar Raspberry Pi logo on the top left and the boot log. After a moment, you are in a very light graphical desktop. There is a giant raspberry on the background (see Figure C-2).

Congratulations, you have now successfully installed Raspbian Linux on your Raspberry Pi!

But wait, there could be more to celebrate! If you haven't installed Linux on a computer before, congratulations again. Now you have installed Linux on a computer. (If this was your first time installing an operating system, our hat is off to you now.)

Now that you have installed Linux on the Raspberry Pi, do you want it on your desktop too? You can get a Free CD image for that from xubuntu (*http://xubuntu.org*) or debian (*http://debian.org*).

Figure C-2. *Booting to the desktop*

Using Raspberry Pi

The graphical desktop makes it easy for you to get started. And you'll get to swim in the deep end of the pool too, as you'll learn to use the famous Linux commands.

The graphical desktop is there to help learning and development. Raspberry Pi is not powerful enough to replace your laptop or desktop. When your project is ready, it's possible to run Raspberry Pi *headless*, without a display, if you want to.

The most important programs on the desktop are the terminal and web browser. You'll use LXTerminal for the command-line interface (CLI). To browse the Web, you use Midori, the light web browser.

If you connected to the wired Internet before booting up the Raspberry Pi, you can start browsing immediately. Double-click the Midori icon on the left of the desktop. When it opens, browse to *http://botbook.com* or *http://google.com*. This verifies that the Internet connection is working. You can also use the browser to copy and paste code examples from *http://botbook.com*, or to copy and paste your own code to your own blog. You do publish your successes, don't you?

To use the command-line interface, double-click the LXTerminal on the desktop. You'll get the prompt, waiting for your commands. It has some text, and at the end, a dollar sign. In the commands shown next, the dollar sign is the prompt, so you don't have to type that character.

Try writing some text. If you live outside the United States, it's likely that some of the characters are wrong. By default, you are using the US keymaps, even though your physical keyboard is different. If you didn't adjust the keyboard with raspi-config earlier, correct the keyboard now. Use your own two-letter ISO country code (fi, fr, de, etc.) in place of the default us.

```
$ setxkbmap us
```

Remember that you don't type the prompt ($), so you only need to type `setxkbmap us` and press Enter.

In the command line, you are always in a directory, which is your *current working directory*. You can ask the computer to print your working directory:

```
$ pwd
/home/pi/
```

As your username in this system is "pi," your home directory is */home/pi/*. This is the only place where a user can permanently store data.

You can change the directory:

```
$ cd /etc/
$ pwd
/etc/
$ cd /home/pi
```

If you ever get lost deep into the beautiful directory tree of Linux, it's easy to get back home. Return to your home directory with *cd /home/pi* or by using the shorthand `cd` all by itself:

```
$ cd
```

You can view the files in the working directory with `ls`:

```
$ ls
Desktop  ocr_pi.png  python_games
```

To create a new file, you can use a text editor:

```
$ nano foo.txt
```

When nano opens, type some text. Press Ctrl-X to save it. Then answer yes to "Save modified buffer?" by typing "Y". Accept the "File Name to Write: foo.txt" prompt by pressing Enter or Return.

You can see your new file in the file listing:

```
$ ls
Desktop  foo.txt  ocr_pi.png  python_games
```

To edit the same file again, type the same command:

```
$ nano foo.txt
```

If you want to see the contents of the file quickly, you can type:

```
$ cat foo.txt
```

When playing with hardware interfaces, you often write just one word into a text file. There is a nice shorthand way of doing that using a *pipe*. First, let's just write some text to the screen:

```
$ echo "Hello, world"
Hello, world
```

Not very surprising yet. But we can take the output of this program "Hello, world" and use it as input for another.

```
$ echo "Hello, world" | tee hello.txt
Hello, world
```

The text "Hello, world" is sent as input to the command `tee` which prints it on the screen and writes it to the specified file (*hello.txt*). The pipe symbol (|) in between the two commands is a character of its own, a tall vertical line.

How do you write the pipe character? In the US keyboard, hold Shift and press the "|\" key. The "|\" key is below the Backspace key and above Return/Enter. In the Finnish keyboard, for example, you can get the pipe character by pressing Alt-Gr and the "<>|" key near the left Shift key.

Check out the contents of the new file you created:

```
$ cat hello.txt
Hello, world
```

This trick will come in handy when you play with the file-based interface to the Raspberry Pi's hardware pins.

Python Programming

You can write programs just like any other text file. Write the first Python program, "Hello, world":

```
$ nano hello.py
```

For the contents of the *hello.py* file, type just this line:

```
print("Hello, world")
```

Press Ctrl-X, type "y," then press Enter to save it.

Now you can run your first Python program with the command *python hello.py*. It should print the text "Hello, world:"

```
$ python hello.py
Hello, world
```

Did you get the text to appear in the window? If so, you just ran some Python code.

Rootly Powers

One of the cornerstones of Linux security and stability is *user separation*. A normal user is only allowed to do things that only affect him. Just try it out:

```
$ raspi-config
Script must be run as root. Try 'sudo raspi-config'
```

You can't change system-wide settings as a normal user.

But luckily, your user "pi" is a member of the "sudo" group, so he can temporarily gain these root powers:

```
$ sudo raspi-config
```

You get the same configuration menu you used earlier to change the password and enable the graphical desktop. Use the Tab and arrow keys to select <Finish> and press Enter (or Return) to get back to command prompt.

Welcome to the Command Line

You have now played with the command-line interface. If you earn your living running Linux servers, you probably knew most of this already. Even on Linux desktops, you might have used the same commands. If not, welcome to the command-line interface!

The most important commands are those that allow you to move around (`ls`, `pwd`, `cd`) and manipulate text files (`nano`, `cat`).

As you'll soon try out yourself, all Linux configuration is in human readable text files. So by adding *sudo*, you can modify anything on the system.

The commands you just learned are surprisingly useful. Most of them work on all Linux system (Debian, Red Hat, Ubuntu, and others), all Unix systems, and even on OS X (which is based on Unix). They have stood the test of time.

Many commands (`ls`, `cd`, `pwd`) have been in use before Linux, before the Web, and even before the Internet protocol suite. So you can likely keep using these commands far into the future.

Take Over an Onboard LED

Hello, hardware! Using this exercise, you will:

- Control the ACT onboard LED
- Meet the GPIO header, the main way to connect hardware to the Raspberry Pi
- Modify the system using virtual files in the */sys/* directory

Install the operating system (see Appendix C) if you haven't done so already. To use the command-line interface, open LXTerminal.

There are five LEDs near the USB connector: ACT, PWR, FDX, LNK, and 100. Could it be possible to take over one of these built-in LEDs?

Luckily, one of the LEDs can be programmatically controlled. The ACT LED is showing SD card activity by default, but we can disable this behavior. Have a look at the current setting:

```
$ cat /sys/class/leds/led0/trigger
none [mmc0]
```

The trigger file you just read shows that currently, the ACT LED (led0) is controlled by mmc0, so it's just showing SD card use. The brackets "[]" show the current selection. The other option is "none," so you will want to select that to control the LED:

```
$ echo "none"|sudo tee /sys/class/leds/led0/trigger
none
```

In Linux, all configuration is in plain-text files. All permanent, system-wide configuration is under */etc/*. And even the ever-changing state of the system can be read and modified trough virtual files in */sys/* and */proc/*. Linux gurus can usually guess the location of the configuration files for any aspect of the system, and then modify the files with just a text editor.

The first part of the command `echo "none"` just prints the word `none`. The second part writes this word to the trigger file. You can see that `none` is now selected, as it's surrounded by square brackets:

```
$ cat /sys/class/leds/led0/trigger
[none] mmc0
```

So now it's time to take control! Write "0" to the "brightness" file to turn lights out:

```
$ echo "0" |sudo tee /sys/class/leds/led0/brightness
```

Look at the Raspberry Pi board to see that the ACT LED is off. Then it's time for lights on:

```
$ echo "1" |sudo tee /sys/class/leds/led0/brightness
```

And you can see the green ACT LED light up.

The ACT light will revert to its default purpose after the next reboot. But if you want to change it back now, you can:

```
$ echo "mmc0"|sudo tee /sys/class/leds/led0/trigger
```

Did you succeed in controlling the light? Well done! You are well on your way to connecting the Raspberry Pi with the physical world.

As you gained control over the onboard LED, you:

- Manipulated the */sys/* virtual filesystem
- Had some hands-on experience with the command-line interface
- Succeeded in controlling some of the Raspberry Pi hardware

Now that you can manipulate the LED built in to the Raspberry Pi, it's time to connect your own.

Hello, GPIO: Connecting an External LED

Hello, external hardware! Soon you'll connect an external LED to your Raspberry Pi. This is the very first external hardware component you add.

Working trough this exercise, you'll control Linux trough virtual text files, and you'll get your first touch to the GPIO pin header.

This is the "Hello, world" for GPIO hardware. If you ever run into problems with any sensors, it's a good idea to repeat this exercise to solve any problems with this simple setup.

Parts

You need the following parts for this project:

- Raspberry Pi
- Female-male jumper wires, black and green

- Breadboard
- 470 Ω resistor (five-band: yellow-violet-black-black-any; four-band: yellow-violet-brown-any)
- An LED

Raspberry Pi has a GPIO pin header for connecting external components. The GPIO pin header has 26 pins (see Figure C-3). Most pins can have multiple functions, and you can choose which one you use. In this experiment, you learn the "out" mode to turn the pin on (3.3 V) and off.

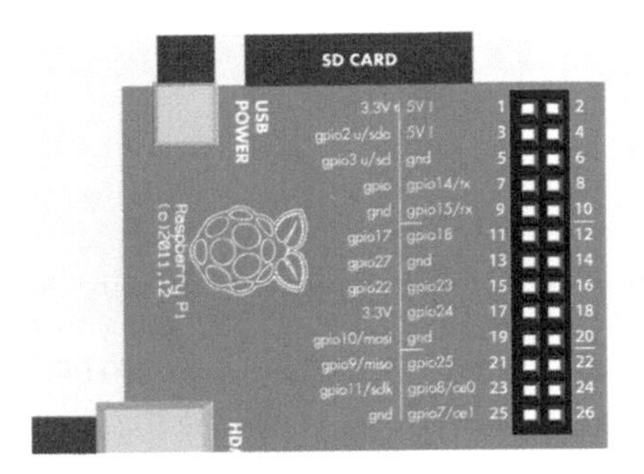

Figure C-3. *The GPIO pin header*

The Raspberry Pi Model B+ features more input/output pins, but the first 26 pins of the B+ header are compatible with earlier models. This means that you can follow the instructions in this book on the B+ without modification.

The circuit (see Figure 4-4) starts from the GPIO 27 pin that you control. When you want to turn on the LED, this pin is +3.3 V, the HIGH of Raspberry Pi. Then it has a resistor-limiting current, to protect both the LED and the more expensive Raspberry Pi. After the resistor, there is an LED in series. Finally, the minus leg of the LED is connected to ground (GND, 0 V).

Building the Circuit

Now it's time to build the circuit (see Figure C-4).

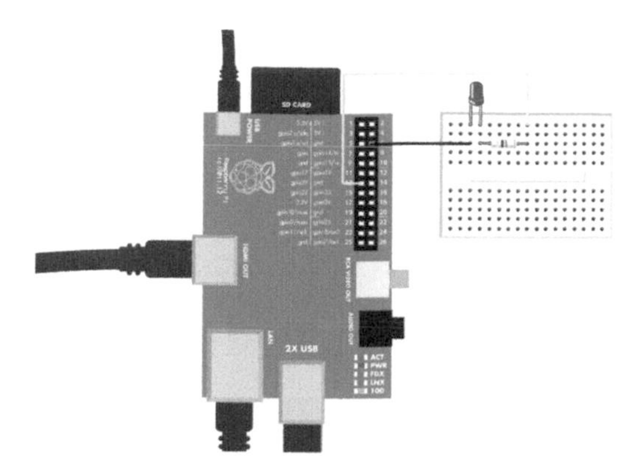

Figure C-4. *Hello, LED*

Unlike Arduino, Raspberry Pi is very sensitive to mistakes. You must be careful to only connect the pins you want to use, so you don't mistakenly short-circuit random pins together.

It is possible to build the circuit while the Raspberry Pi is on. Of course, mistakenly shorting two pins could then immediately break your Raspberry. The safer option is to shut down Raspberry Pi before connecting the wires. You can find the shutdown in the main menu, the menu on the bottom left of the desktop (Main Menu → Logout → Shutdown). After the shutdown, disconnect the powering USB cable.

Find the correct pins on the GPIO header. The pin number one is marked with a small white box on the board. The numbering diagram (see Figure C-3) shows both the human-friendly name (GPIO 27) and the running number of the pin (13). The pins are:

- GPIO 27 is pin number 13
- GND is pin number 14

Connect the female end of the green jumper wire to GPIO 27 and push the male end to a free row in the breadboard. Push one leg of the resistor to the same row with the jumper wire.

Push the free leg of the resistor to a free row on the breadboard. Push the plus leg of the LED to this row.

Push the male end of a black jumper wire into the row with the LED minus leg. Connect the female end of the black jumper wire to the GND pin on the GPIO header.

Verify that you got the same kind of connections as in Figure C-4. After you have double-checked the connections, power up the Raspberry Pi if it was off while making the connections.

Commands to Control GPIO

Open LXTerminal by double-clicking the LXTerminal icon on the desktop. Then take control of your GPIO pin.

In */sys/* you have text files that have information about the state of the system. These files are virtual, so they only exist in computer's memory.

Export the GPIO 27 pin, so we can use it:

```
$ echo 27|sudo tee /sys/class/gpio/export
```

This created a new folder *gpio27*. Turn the pin to "out" mode, so you can control it:

```
$ echo out|sudo tee /sys/class/gpio/gpio27/direction
```

Turn the pin on (+3.3 V, HIGH):

```
$ echo 1|sudo tee /sys/class/gpio/gpio27/value
```

The LED lights up. Did you get it to light up? Well done! You can now connect some hardware to your Raspberry Pi!

You can shut down the light, too:

```
$ echo 0|sudo tee /sys/class/gpio/gpio27/value
```

Troubleshooting

The LED is not lit, even though all commands seem to run. Check the LED polarity. The long plus leg goes to the plus side, the side of the GPIO 27 pin. On the minus side of the LED, there is a flat cut in the plastic. The minus side goes toward the black wire, GND. If this doesn't help, check the connections.

Hello, GPIO world! As you have learned, every project starts with a "Hello, world." If you ever run into problems with sensors, it could be a good idea to build and run this "Hello, GPIO" project again. That way, you get to solve the simple problems with a simple setup.

For us, "Hello, world" is usually the hardest project. It's not just an LED! Your "Hello, GPIO world" proves that:

- You have successfully installed Linux on your Raspberry Pi.
- You can use the command line.

- You can manipulate the state of the GPIO pins.
- You have obtained components that work with the Raspberry Pi.
- You have built a circuit that works with the Raspberry Pi (and did not fry it).

If there is a time for a break, a successful "Hello, world" might be it. Reward you brain with a cup of coffee, some jogging, or staring out of the window for a while.

Using GPIO Without Root

By default, you need to invoke root user privileges on your Raspberry Pi to access the GPIO pins in your code. This section shows you how to configure your Raspberry Pi to avoid that. You only need to go through this process once after you install your Raspberry Pi operating system.

Avoiding root privileges will make the system more secure and more stable. For example, think about a program that serves sensor data to the Web. Would you run a program that strangers can connect to as root?

In Linux, devices attached to your system are controlled by *udev*. Udev is a rule-based system that can run scripts when devices are plugged in.

Linux lets you control GPIO pins by manipulating files in */sys/class/gpio/*. By default, these files are owned by the user "root" and the group "root," which is why you can't change them without invoking root's superuser privileges. In this section, you'll see how to write a udev rule to change the group to "dialout." You'll then allow that group to read and write the files under */sys/class/gpio/*. Finally, you'll make the folders' group *sticky*, so that any newly created files and folders under it will also be owned by the "dialout" group.

All system-wide configuration in Linux is under */etc/*. Not surprisingly, udev configuration is in */etc/udev/*. Power up your Raspberry Pi, and open the terminal (LXTerminal). Next, open an editor with the `sudoedit` command so you can create a new rule file (don't type the `$`; that indicates the shell prompt that you see in the terminal window):

```
$ sudoedit /etc/udev/rules.d/88-gpio-without-root.rules
```

Add the text shown in Example C-1 to the file. Be sure to type each line as shown (don't type the numeric symbols; those are there to explain to you what is going on in this file). Udev rules are very sensitive to typos.

Want to save typing? Download the code (*http://getstarted.botbook.com*), unzip it, and copy to */home/pi* on your Raspberry Pi SD card.

If you don't have permission to read the directory on Raspberry Pi, use `sudo chown -R pi.pi code/`.

Example C-1. 88-gpio-without-root.rules

```
# /etc/udev/rules.d/88-gpio-without-root.rules - GPIO without root on Raspberry
Pi  # ❶
# Copyright http://BotBook.com
# ❷
SUBSYSTEM=="gpio", RUN+="/bin/chown -R root.dialout /sys/class/gpio/"
SUBSYSTEM=="gpio", RUN+="/bin/chown -R root.dialout /sys/devices/virtual/gpio/"
# ❸
SUBSYSTEM=="gpio", RUN+="/bin/chmod g+s /sys/class/gpio/"
SUBSYSTEM=="gpio", RUN+="/bin/chmod g+s /sys/devices/virtual/gpio/"
# ❹
SUBSYSTEM=="gpio", RUN+="/bin/chmod -R ug+rw /sys/class/gpio/"
SUBSYSTEM=="gpio", RUN+="/bin/chmod -R ug+rw /sys/devices/virtual/gpio/"
```

❶ This comment explains the purpose of the file.

❷ Sets the owner of the two directories to be root, and the group to be dialout.

❸ Sets the *sticky bit* flag on these two directories.

❹ Configures the permissions on the directories to give members of the dialout group read and write permission.

The rules are processed in numeric order, but this is probably the only rule affecting the GPIO directories, so the number does not matter. In Morse code (CW), 88 is short for hugs and kisses. We prefer it over the often-picked number 99, which means "get lost".

To avoid typing and inevitable typos, you can download a copy of the *88-gpio-without-root.rules* file from *http://botbook.com* and copy it in place with *sudo cp 88-gpio-without-root.rules /etc/udev/rules.d/*. To edit the file, you can use `sudoedit`, which opens the file as root in the `nano` text editor:

```
$ sudoedit /etc/udev/rules.d/88-gpio-without-root.rules
```

Save the file (Ctrl-X, press y, and then press Enter).

To use your new rules, restart the udev daemon and trigger your new rule with these commands:

```
$ sudo service udev restart
$ sudo udevadm trigger --subsystem-match=gpio
```

Next, check whether the ownership on the files was set correctly:

```
$ ls -lR /sys/class/gpio/
```

The listing should mention the "dialout" group many times. The parameter `-l` means to display a long listing (with owner, group, and permissions), and `-R` means recursively list directory contents, too. If you don't see dialout in the listing, restart your Raspberry Pi and check again with the `ls -lR /sys/class/gpio/` command. If it still doesn't work, go through the steps to configure the udev rule again.

Once you have the new udev rule in place, you won't need to go through the preceding steps again unless you reinstall the operating system on your Raspberry Pi.

D/Bill of Materials

This section lists the parts you'll need to complete the projects in this book.

Chapter 1

The parts for "Project 1: Photoresistor to Measure Light" on page 3 are as follows:

- Photoresistor
- 5 mm red LED (different LEDs will work differently with this circuit; later, you'll learn a more sophisticated way to fade LEDs)
- 470 Ω resistor (four-band resistor: yellow-violet-brown; five-band resistor: yellow-violet-black-black; the last band will vary depending on the resistor's tolerance)
- Breadboard
- 9 V battery clip
- 9 V battery

Chapter 2

The parts for "Project 2: A Simple Switch" on page 9 are as follows:

- A switch
- A wire with alligator clips
- Two 1.5 V batteries
- Battery holder with wire leads
- LED
- 470 Ω resistor (four-band resistor: yellow-violet-brown; five-band resistor: yellow-violet-black-black; the last band will vary depending on the resistor's tolerance)
- Breadboard

All of these except the 9 V battery and 470 Ω resistor are available in the Maker Shed Mintronics: Survival Pack (*http://bit.ly/mintron-sp*), part number MSTIN2. You can use two of the 220 Ω resistors in series or one 1 kΩ resistor in place of the 470 Ω resistor; both of these are available from electronics retailers such as RadioShack.

The parts for "Project 3: Buzzer Volume Control" on page 12 are as follows:

- DC piezo buzzer (Maker Shed part number MSPT01)
- 10 K potentiometer
- 9 V battery
- Battery clip
- Breadboard

All of these except the 9 V battery and piezo buzzer are available in the Maker Shed Mintronics: Survival Pack (*http://bit.ly/mintron-sp*), part number MSTIN2.

The parts for "Project 4: Hall Effect" on page 14 are as follows:

- DC piezo buzzer
- 9 V battery
- Battery clip
- Breadboard
- NJK-5002A Hall effect switch

The parts for "Project 5: Firefly" on page 17 are as follows:

- Two BC547 transistors
- Light-dependent resistor (LDR)
- 470 Ω resistor (four-band resistor: yellow-violet-brown; five-band resistor: yellow-violet-black-black; the last band will vary depending on the resistor's tolerance)
- 10 kΩ resistor (four-band resistor: brown-black-orange; five-band resistor: brown-black-black-red; the last band will vary depending on the resistor's tolerance)
- 5 mm red LED
- 9 V battery

- 9 V battery clip
- Breadboard
- CA555E timer integrated circuit
- 10 kΩ potentiometer
- 100 uF capacitor

All of these parts, except the 9 V battery, 470 Ω resistor, and BC547, are available in the Maker Shed Mintronics: Survival Pack (*http://bit.ly/mintron-sp*), part number MSTIN2. You can use two of the 220 Ω resistors in series or one 1 kΩ resistor in place of the 470 Ω resistor; these are both available from electronics retailers such as RadioShack.

Chapter 3

Chapter 3 requires an Arduino to interface with your sensors. If you need to buy one, purchase an Arduino Uno or an Arduino Leonardo.

The parts for "Project 6: Momentary Push-Button and Pull-Up Resistors" on page 34 are as follows:

- Momentary push button
- Arduino Uno
- Jumper wires
- Breadboard

The parts for "Project 7: Infrared Proximity to Detect Objects" on page 40 are as follows:

- Infrared sensor switch
- Arduino Uno
- Jumper wires

The parts for "Project 8: Rotation (Pot)" on page 43 are as follows:

- Potentiometer (around 10 kΩ recommended)
- Arduino Uno
- Jumper wires
- Breadboard

The parts for "Project 9: Photoresistor to Measure Light" on page 47 are as follows:

- Photoresistor (10 K recommended)
- 10 kΩ resistor (four-band resistor: brown-black-orange; five-band resistor: brown-black-black-red; the fourth or fifth band will vary depending on the resistor's tolerance)
- Arduino Uno
- Jumper wires
- Breadboard

The parts for "Project 10: FlexiForce to Measure Pressure" on page 49 are as follows:

- Force-sensitive resistor (FSR)
- 1 MΩ resistor (four-band resistor: brown-black-green; five-band resistor: Brown-black-black-yellow; the fourth or fifth band will vary depending on the resistor's tolerance)
- Arduino Uno
- Jumper wires
- Breadboard

The parts for "Project 11: Measuring Temperature (LM35)" on page 52 are as follows:

- LM35 temperature sensor
- Arduino UNO
- Jumper wires
- Breadboard

The parts for "Project 12: Ultrasonic Distance Measuring (HC-SR04)" on page 56 are as follows:

- HC-SR04 ultrasonic distance sensor
- Arduino UNO
- Breadboard
- Jumper wires

Chapter 4

The parts for "Project 13: Momentary Push Button" on page 64 are as follows:

- Momentary Push Button
- Raspberry Pi
- Female-male jumper wires (Maker Shed part #MKKN5), black and green
- A breadboard

The parts for "Project 14: Blink an LED with Python" on page 69 are as follows:

- Raspberry Pi
- Female-male jumper wires (Maker Shed part #MKKN5), black and green
- An LED
- 470 Ω Resistor (four-band resistor: yellow-violet-brown; five-band resistor: yellow-violet-black-black; the last band will vary depending on the resistor's tolerance)
- A breadboard

The parts for "Project 15: Adjustable Infrared Switch" on page 73 are as follows:

- Infrared sensor switch
- Two 1 kΩ resistors (five band: brown-black-red-brown-any, four band: brown-black-red-any)
- Raspberry Pi

The parts for "Project 16: Potentiometer to Measure Rotation" on page 77 are as follows:

- Potentiometer
- MCP3002 (analog-to-digital integrated circuit)
- Raspberry Pi
- Jumper wires (male-to-female and male-to-male, which are Maker Shed parts #MKKN5 and #MKSEEED3)
- Breadboard

The parts for "Project 17: Photoresistor" on page 82 are as follows:

- Photoresistor (10 kΩ suggested)
- Resistor (10 kΩ, or choose a value that matches the photoresistor)
- MCP3002
- Raspberry Pi

- Jumper wires (male-to-female and male-to-male, which are Maker Shed parts #MKKN5 and #MKSEEED3)
- Breadboard

The parts for "Project 18: FlexiForce" on page 85 are as follows:

- Flexiforce (11 kg, 25 pound version recommended)
- Resistor (1 MΩ, five band: brown-black-black-yellow-any, four band: brown-black-green-any)
- MCP3002
- Raspberry Pi
- Jumper wires (male-to-female and male-to-male, which are Maker Shed parts #MKKN5 and #MKSEEED3)
- Breadboard

The parts for "Project 19: Temperature Measurements (LM35)" on page 86 are as follows:

- LM35 temperature sensor
- MCP3002
- Raspberry Pi
- Jumper Wires (male-to-female and male-to-male, which are Maker Shed parts #MKKN5 and #MKSEEED3)
- Breadboard

The parts for "Project 20: Ultrasonic Distance" on page 89 are as follows:

- HC-SR04 sensor
- Two 10 KΩ resistors (four band: brown-black-orange-any, five band: brown-black-black-red-any)
- Raspberry Pi
- Jumper wires (male-to-female and male-to-male, which are Maker Shed parts #MKKN5 and #MKSEEED3)
- Breadboard

Index

Symbols

About the Authors

Kimmo Karvinen works as a chief technology officer for a hardware manufacturer that specializes in integrated AV and security systems. Before that, he worked as a marketing communications project leader and as a creative director and partner in an advertisement agency. Kimmo's education includes a Master of Arts.

Tero Karvinen teaches Linux and embedded systems in Haaga-Helia University of Applied Sciences, where his work has also included curriculum development and research in wireless networking. He previously worked as a CEO of a small advertisement agency. Tero's education includes a Master of Science in Economics.

The cover and body font is Benton Sans, the heading font is Serifa, and the code font is Bitstream Vera Sans Mono.

Lightning Source UK Ltd.
Milton Keynes UK
UKOW06f0330130914

238423UK00002B/3/P